GLOBAL MIGRATION

THE BASICS

Migration is a politically sensitive topic and an important aspect of contentious debates about social and cultural diversity, economic stability, terrorism, globalization, and nationalism. *Global Migration: The Basics* provides an accessible and balanced introduction to this important area of study, providing a detailed examination of current scholarship, demographic patterns, and public policy debates to discuss:

- the history and geography of global migration
- the role of migrants in society
- the impact of migrants on the economy and the political system
- the policy challenges that need to be addressed in a rapidly changing world.

With a series of engaging case studies and suggestions for further reading, *Global Migration: The Basics* exposes readers to the underlying causes and consequences of migration. This book is an essential read for any students wishing to look beyond the rhetoric and consider the facts about migration.

Bernadette Hanlon is Assistant Professor in the City and Regional Planning Program in the Knowlton School of Architecture at The Ohio State University, USA.

Thomas J. Vicino is Associate Professor of Political Science at Northeastern University in Boston, where he is also affiliated with the School of Public Policy and Urban Affairs.

THE BASICS

ACTING
BELLA MERLIN

AMERICAN PHILOSOPHY
NANCY STANLICK

ANCIENT NEAR EAST
DANIEL C. SNELL

ANTHROPOLOGY
PETER METCALF

ARCHAEOLOGY (SECOND EDITION)
CLIVE GAMBLE

ART HISTORY
GRANT POOKE AND DIANA NEWALL

ARTIFICIAL INTELLIGENCE
KEVIN WARWICK

THE BIBLE
JOHN BARTON

BIOETHICS
ALASTAIR V. CAMPBELL

BUDDHISM
CATHY CANTWELL

THE CITY
KEVIN ARCHER

CONTEMPORARY LITERATURE
SUMAN GUPTA

CRIMINAL LAW
JONATHAN HERRING

CRIMINOLOGY (SECOND EDITION)
SANDRA WALKLATE

DANCE STUDIES
JO BUTTERWORTH

EASTERN PHILOSOPHY
VICTORIA S. HARRISON

ECONOMICS (SECOND EDITION)
TONY CLEAVER

EDUCATION
KAY WOOD

EUROPEAN UNION (SECOND EDITION)
ALEX WARLEIGH-LACK

EVOLUTION
SHERRIE LYONS

FILM STUDIES (SECOND EDITION)
AMY VILLAREJO

FINANCE (SECOND EDITION)
ERIK BANKS

FREE WILL
MEGHAN GRIFFITH

HUMAN GENETICS
RICKI LEWIS

HUMAN GEOGRAPHY
ANDREW JONES

INTERNATIONAL RELATIONS
PETER SUTCH AND JUANITA ELIAS

ISLAM (SECOND EDITION)
COLIN TURNER

JOURNALISM STUDIES
MARTIN CONBOY

JUDAISM
JACOB NEUSNER

LANGUAGE (SECOND EDITION)
R.L. TRASK

LAW
GARY SLAPPER AND DAVID KELLY

LITERARY THEORY (SECOND EDITION)
HANS BERTENS

LOGIC
JC BEALL

MANAGEMENT
MORGEN WITZEL

MARKETING (SECOND EDITION)
KARL MOORE AND NIKETH PAREEK

MEDIA STUDIES
JULIAN MCDOUGALL

THE OLYMPICS
ANDY MIAH AND BEATRIZ GARCIA

PHILOSOPHY (FIFTH EDITION)
NIGEL WARBURTON

PHYSICAL GEOGRAPHY
JOSEPH HOLDEN

POETRY (SECOND EDITION)
JEFFREY WAINWRIGHT

POLITICS (FOURTH EDITION)
STEPHEN TANSEY AND NIGEL JACKSON

PUBLIC RELATIONS
RON SMITH

THE QUR'AN
MASSIMO CAMPANINI

RACE AND ETHNICITY
PETER KIVISTO AND PAUL R. CROLL

RELIGION (SECOND EDITION)
MALORY NYE

RELIGION AND SCIENCE
PHILIP CLAYTON

RESEARCH METHODS
NICHOLAS WALLIMAN

ROMAN CATHOLICISM
MICHAEL WALSH

SEMIOTICS (SECOND EDITION)
DANIEL CHANDLER

SHAKESPEARE (THIRD EDITION)
SEAN MCEVOY

SOCIAL WORK
MARK DOEL

SOCIOLOGY
KEN PLUMMER

SPECIAL EDUCATIONAL NEEDS
JANICE WEARMOUTH

TELEVISION STUDIES
TOBY MILLER

TERRORISM
JAMES LUTZ AND BRENDA LUTZ

THEATRE STUDIES (SECOND EDITION)
ROBERT LEACH

WOMEN'S STUDIES
BONNIE SMITH

WORLD HISTORY
PETER N. STEARNS

GLOBAL MIGRATION

THE BASICS

Bernadette Hanlon and Thomas J. Vicino

LONDON AND NEW YORK

First published 2014
by Routledge
2 Park Square, Milton Park, Abingdon, Oxon OX14 4RN

and by Routledge
711 Third Avenue, New York, NY 10017

Routledge is an imprint of the Taylor & Francis Group, an informa business

British Library Cataloguing in Publication Data
A catalogue record for this book is available from the British Library

Library of Congress Cataloging in Publication Data
Hanlon, Bernadette, 1969–
Global migration : the basics / Bernadette Hanlon, Thomas J. Vicino.
pages cm. – (The basics)
1. Human beings – Migrations. 2. Emigration and immigration.
3. Human geography. I. Vicino, Thomas J. II. Title.
GN370.H36 2014
304.8 – dc23
2013034112

ISBN: 978-0-415-53385-0 (hbk)
ISBN: 978-0-415-53386-7 (pbk)
ISBN: 978-1-315-88246-8 (ebk)

Typeset in Bembo and Scala Sans
by Taylor & Francis Books

Printed and bound by CPI Group (UK) Ltd,
Croydon, CR0 4YY

For John Rennie Short

CONTENTS

LIST OF ILLUSTRATIONS

FIGURES

TABLES

PREFACE

As we finished writing this book, the City of Boston was still recovering from the bombings at the Boston Marathon on April 15, 2013. The two primary suspects in the bombing were identified as brothers who migrated to the United States as refugees in 2002. Two days after the bombing, the two suspects allegedly carjacked a vehicle. The owner of the vehicle was held hostage, but he later escaped when the suspects stopped at a gas station. The hostage was Danny, a Chinese migrant living in the Boston metropolitan area. After his escape from the bombing suspects, he ran to a gas station and asked a clerk, an immigrant from Egypt, to call the police. Migrants, like all people, can be good or evil, brave or cowardly, can demonstrate true citizenship or harbor hatred. In most cases, like most people, they live regular lives. Yet in a few cases, socio-political tensions surrounding migration can lead to destructive results.

This story demonstrates that cities are indeed diverse, global places. Cities are places that have flourished, declined, and rebounded as the patterns of migration ebb and flow. The people of cities shape the experiences of all those who call them home. As scholars of urbanism, we have a strong interest in cities. Our interest spills over into our own lives: we have lived in cities, attended schools in cities, worked in cities, and enjoyed cities. Our own perspectives on the city have shaped our view of migration. We live in the largest nation of immigrants: the United States. And we share an immigrant legacy. We share a heritage from Ireland and Italy—countries whose people migrated in large numbers to the United States. Our relatives live in different parts of the world. In this book, we hope to share our excitement about migration and its consequences.

In joining other books in *The Basics* series, we offer a broad overview of the study of migration. In this volume, we provide a glimpse into the historic and contemporary issues of migration across the globe through a thorough synthesis of much of the academic literature and data on migration. We present an introduction to the study of migration through a variety of disciplinary lenses. Grounded in the social sciences, we draw on sociology, economics, geography, and politics to shape our understanding of complexities of where, when, how, and why people migrate. We assume that the reader is studying migration for the first time. The book is not designed to be an exhaustive examination of migration; rather, we attempt to distill from a vast literature the central questions and key concepts on the topic. Ultimately, our goal is to capture the reader's interest and to bring alive the topic of migration through a series of case studies. The chapters are structured thematically, and they are supplemented with a guide to further reading and a sample of useful resources. Much like any other area in the socio-political realm, migration can be a controversial and sensitive topic of discussion. We urge the reader to remember as you read this book that migrants are people, just like you, looking to provide for themselves, their families, and their communities.

In writing *Global Migration: The Basics*, we consulted a number of exemplary books on the topic. Among others, these included: *The Age of Migration* by Stephen Castles and Mark Miller; *International Migration: A Very Short Introduction* by Khalid Koser; *Migration Theory: Talking Across Disciplines* edited by Caroline B. Brettell and James F. Hollifield; *Global Migration Governance* edited by Alexander Betts; and a valuable resource, *The Encyclopedia of Global Human Migration* edited by Immanuel Ness and Peter Bellwood. We recommend these primary works to anyone interested in further investigating the topic of migration. We also referred to a number of key academic journals. These included: *International*

Migration; *International Migration Review*; the *Journal of Ethnic and Migration Studies*; and the *International Social Science Journal*. In synthesizing the literature for this book, we are indebted to the many scholars who have written on the topic of international migration.

We would also like to acknowledge a number of people individually, but before doing so, together we would especially like to acknowledge one of our finest mentors, John Rennie Short, a consummate scholar, to whom this book is dedicated. We have collaborated with John for nearly a decade, and we have learned a great deal from him. We are forever grateful for these shared experiences.

Bernadette Hanlon would like to thank her colleagues in the City and Regional Planning Program at the Ohio State University (OSU) for their support in this endeavor, including Jennifer Evans-Cowley, Rachel Kleit, Maria Manta-Conroy, Hazel Morrow-Jones, Kenneth Pearlman, Charisma Acey, Kyle Ezell, Gulsah Akar, Jesus Lara, Jack Nasar, Philip Viton, and Jean-Michel Guldman. I would also like to acknowledge colleagues in OSU's Knowlton School of Architecture, including Michael Cadwell and Kay Bea Jones for supporting my intellectual pursuits. Last, I would also like to thank my family for their comfort and support. I am especially grateful to Kerry McCarthy for her encouragement when writing this book.

Thomas Vicino would like to thank his colleagues at Northeastern University for their continued support of his work. In particular, fruitful discussions with Amílcar Barreto, Denise Garcia, Denise Horn, Berna Turam, and Liza Weinstein stimulated my thoughts on migration. My colleagues in the Department of Political Science, the School of Public Policy and Urban Affairs, and the College of Social Sciences and Humanities, including Mitchell Orenstein, Joan Fitzgerald, and Uta Poiger, provided an exciting intellectual home for global research. My experiences in many cities in Brazil further enlightened my understanding of the complexities of migration. Last, I am grateful for the encouragement that my family always provides. I am appreciative of the support from Tom Armstrong, Jessica Emory, Zachary Haney, Nora Rasanen, Larry Pacific, Jana Vergados, and, finally, from Charles Galantini.

Finally, we would jointly like to thank the editorial team at Routledge, including Siobhan Poole, Iram Satti, and Andy Humphries, for their patience, support, and guidance in the production of this book.

Bernadette Hanlon
Columbus, Ohio
Thomas J. Vicino
Boston, Massachusetts
June 2013

1

MIGRATION: WHAT IT MEANS AND WHY IT HAPPENS

Throughout history, human populations have moved from place to place. As hunters, gatherers, and nomads, we have moved in search of food and shelter. Fleeing famine, natural disasters, and potential aggression from other humans, we sought out new territories. Migration has been central to population dispersal across the world, and chronicles of population movements are important parts of our history and psychology. In the Bible, Noah migrated to avoid the floods; Abraham led his people to the "promised land"; and Mary and Joseph fled with Jesus from Egypt, escaping the terror of a tyrant. Many stories in Judeo-Christian belief are stories of people migrating to different places. Similarly in the Hellenic tradition, Homer's epic poem describes the wanderings of Odysseus, a classic tale of migration. In ancient history, great migration events are connected with Roman expansion, Viking invasions, the Crusades, and the extension of the Inca Empire. There are similarly great conquerors associated with the invasion into new lands—think of Attila the Hun or Genghis Khan. The stories of migration and migrants are legendary and varied. Indeed, the migration process is as old as human civilization.

In this book, we provide a concise but comprehensive introduction to the topic of migration across the world. We focus on migration in more contemporary times than in the pre-modern legends and events. Most major movements of people have occurred over the last few centuries (Koser, 2007). So, although migration has a very long history, we will examine migration and migration processes during the present-day period, with reflection on a recent history of migration.

The purpose of this book is to examine—from varied interdisciplinary perspectives and across different geographies—the social, economic, historical, and political issues around the movement of people. The book synthesizes numerous bodies of literature on migration to convey the essential knowledge base and latest thinking on the topic—all, we hope, in a cohesive voice, usable format, and accessible style. We include a series of case studies scattered throughout the book. At the end of each chapter we provide the reader with examples of books, journal articles, and websites that offer further insights into the topics covered in that chapter.

In this introductory chapter and throughout the book, we provide statistics on global migration. The United Nations Statistics Division and the International Organization for Migration (IOM) represent two valuable sources for migration statistics. The IOM is probably the most comprehensive resource for determining the extent of migration between countries. Established in 1951, at a time of major upheaval across Europe, the IOM began with a mandate to help in the resettlement of the millions of people displaced by the chaos of World War II. This organization has morphed and broadened its scope to, among other things, help governments and non-government organizations across the world manage migration. The IOM is an entity outside of the institutional framework of the United Nations and, while it serves governments in their efforts to manage migration in a way consistent with international law and human rights, this organization does not act with any international regulatory power. As we discuss in Chapter 5, there is no international migratory policy-making entity or structure that directly governs global migration. However, as part of its efforts, the IOM does consolidate data to facilitate data collection of migration patterns. The United Nations Population Division also participates in these data collection efforts. In particular, in Chapter 2, we conduct a nuanced analysis of migration trends using data from these two organizations as well as other sources, and in doing so we aim to identify the various patterns of migration both over time and across different geographies.

In this introductory chapter, we explain what we mean by migration and include an examination of the different forms of migration and the different types of migrants we witness today. Then, we identify the major theories to explain why global migration occurs. We conclude this introductory chapter with a discussion of how this book proceeds, and identify the main topics covered in each of the remaining chapters. For now, though, let us begin by identifying different types of migration and different types of migrants.

DIFFERENT TYPES OF MIGRATION

To begin the study of global migration, it is necessary to distinguish between different types of migration and, in the next section, the different types of migrants. Let us begin with a discussion of two important types of migration. First, there is internal migration. This refers to the movement and relocation of people inside one country. People move from one part of the country to another to establish a new residence. With the rise of the global economy and the growing importance of urban centers, the general trend in internal migration is the movement of people from rural areas within their nation to cities of their nation, often in great numbers (Saunders, 2010). For instance, in 2009, an estimated 145 million people migrated from rural villages in China to large fast-growing cities in China. Internal migration has dramatically reshaped the urban footprint of many of the world's largest developing cities, from China, to Brazil, to Indonesia.

The world's population is generally urbanizing, and this is expected to continue throughout the twenty-first century. According to the *U.N. World Urbanization Prospects: The 2011 Revision* report, more than half of the world's population now live in urban areas. By 2050, estimates suggest that 86 percent of the population of developed regions and 67 percent of the population of less developed regions will be urban dwellers. At the same time, the rural population is expected to decline over time. Between the years 2011 and 2050, the rural population is expected to fall from 3.1 billion to 2.9 billion people. Internal migration will contribute to these changes as people from a nation's rural towns and villages relocate to that nation's cities. Internal migration is part of the urbanization process.

A second type of migration is the movement and relocation of people from one nation to another. This is external migration, or international migration, which we refer to as global migration. This type of migration is the focus of this book. As with internal migration, external or global migration contributes to the growth of urban centers in many countries. As we will explore in Chapter 4, global migrants tend to seek out new opportunities in cities. According to the United Nations (UN), in 2010, there were some 214 million people—3.1 percent of the world's population—living outside the country where they were born. In 2010, most of these global migrants lived in developed regions of the world. According to UN statistics, about 128 million, or close to 60 percent, of all global migrants lived in North America, Australia, Japan, New Zealand, and Europe. The remaining 86 million migrants lived in the less developed regions of Africa, Asia (excluding Japan), Latin America, and the Caribbean.

Nuances of the historic and geographic trends in global migration will be dealt with in more detail in Chapter 2, but it is worth noting that the general tendency at the current time is the movement of global migrants from less developed regions to more developed regions of the world. This pattern of movement from the South to the North demonstrates a shift in population from poorer to wealthier countries. In this chapter, we discuss some of the major theories that explain why migration occurs. Inequality among nations is explained as a contributing factor.

SOUTH–SOUTH MIGRATION

It is important to recognize that there are movements of people within less developed regions of the world. Migrants move from one country to another within the Global South. For instance, there are an increasing number of Asian migrants living in the Gulf States, and there are many people originally from a variety of sub-Saharan African states that moved to live in South Africa. According to a World Bank report (Ratha and Shaw, 2007), South–South migration is almost as common as South–North migration. This report states that in 2010, approximately 73 million migrants who were born in the South were residing in the South. The growth rate of migrant stock in less developed regions surpassed the growth rate for developed regions for the first time ever during the period from 2005 to 2010. While the reasons for increasing South–South migration are complex, there are indications that flight from ecological disasters, civil unrest, and conflict are contributing factors. Proximity is also important. South–South migrants are poor, and therefore they are restricted to traveling by land. Many people who migrate from a country in the South move to an adjacent or nearby country within that region (Ratha and Shaw, 2007).

THE DIFFERENT TYPES OF MIGRANTS

The UN defines a migrant as a person who resides outside of their country of origin for a period of at least one year. It is quite difficult to assess how many global migrants there actually are at any given time, and so, while we offer some statistics throughout this book, we offer them with some caution. For example, in many cases the official statistics on migration do not include people who have entered a foreign country without the required legal documentation. In other words, official statistics might not include illegal or irregular migrants. In addition, as Castles and Miller (2009) note in their classic book, *The Age of Migration*, terminology is important when considering migration data. The United States collects data on the "foreign-born" population and includes among this group those who obtained citizenship status. In Europe, the term "foreign national" is used when measuring migration and excludes those who have chosen to become nationals of the receiving country. This is just one example of how it is important to be aware of definitions and discrepancies when comparing statistics across different countries.

Within the category of migrants, there is often a distinction made between voluntary migrants and involuntary migrants. Let us first discuss involuntary migrants. As the terminology suggests, these are people who are forced to leave their country or a region within their country because of some sort of conflict, persecution, or environmental disaster such as a drought, flood, or famine.

Involuntary migration is often referred to as forced migration, and we discuss this in more depth in Chapter 5. Often, the move is for political reasons. As a result, involuntary migrants are sometimes referred to as political migrants. In some cases, involuntary migrants become refugees and asylum seekers. In 1951, the UN Convention relating to the Status of Refugees defined a refugee as someone who "owing to a well-founded fear of being persecuted for reasons of race, religion, nationality, membership of a particular social group or political opinion, is outside the country of his [or her] nationality, and is unable to or, owing to such fear, is unwilling to avail himself of the protection of that country." An asylum seeker is a person who has applied for international protection for similar reasons of fear. We discuss the policies and status of refugees and asylum seekers in more detail in Chapter 5. According to the UN, there were 16 million refugees and asylum seekers in 2009.

Some forced or involuntary migrants remain within their country. These migrants are referred to as Internally Displaced People (IDP). They move for such reasons as an environmental disaster, or they are forced to move because of internal conflict and war in their own country. These people are a very vulnerable group who remain under the legal protection of their own government even though that government might be the cause of their movement and distress, or may not be in a position to help them. According to the UN, there were estimated to be about 26.4 million IDP in 2011.

With the rise of environmental problems associated with climate change, there is now a relatively new category of migrants, referred to as environmental migrants. These are people who leave their country or a region within their country because of sudden or long-term environmental events such as drought, desertification, flooding, sea-level rise, hurricanes, tsunami, or monsoons. Environmental migrants sometimes move to another country and sometimes become internally displaced. In many cases, environmental migrants are involuntary migrants because of the suddenness of the environmental event that forces them to move. In the case of environmental change that evolves more slowly, such as sea-level rise, the movement of environmental migrants is more "voluntary." As changes to the built and natural environment occur, these migrants recognize that remaining in certain areas might prove unsafe or economically unwise in the near future. Determining environmental migration is a challenge because it is often difficult to know if the migration event in question is due to environmental or other factors. We examine in more detail the relationship between the environment and migration in Chapter 6.

This brings us to another category of migrants, the voluntary migrant. As the term suggests, these migrants are thought to move voluntarily because economic conditions and living standards in their country or region of origin are not as desirable as in another country or region. In other words, voluntary migrants move to find work and better job opportunities. It should be noted at this point that voluntary migration is a contested term. The notion of voluntary migration suggests that individuals with perfect knowledge make rational decisions to move for better living conditions but, as we explore later, there are complications to this perception. The reasons that people move from their country of origin to another country are varied and complex. Despite this, however, in broad terms, voluntary migrants move largely for economic reasons. Thus, they are referred to as economic or more specifically labor migrants.

Labor migrants generally fall into two categories: high-skilled and low-skilled labor migrants. The high-skilled migrant labor force includes engineers, scholars, experts in Information Technology (IT), artists, technocrats from governments and non-governmental organizations, graduate students, and skilled entrepreneurs. As members of a highly skilled mobile elite, these migrants are sometimes lured from less developed countries to work in more advanced economies where wages are higher than in their home countries and benefits such as health insurance and generous pensions are much more attractive. However, there is concern about the problem of "brain-drain," as technically skilled and educated workers migrate from less developed to more developed countries. We examine in more detail the impact of high-skilled labor migrants on both receiving and sending economies in Chapter 4.

In more recent years, immigration policies in many advanced economies have been designed and implemented to accept and attract high-skilled rather than low-skilled workers. Those migrants who enter a country without the required legal documentation, who forge these documents, or overstay their visas or work permits, tend to be unskilled or poorly skilled labor migrants. Migrants that settle in a country without the correct legal status are referred to as illegal or irregular migrants. In this book, we use the term irregular migration since, like Koser (2007), we believe the term "illegal" inaccurately suggests migrants are criminals when most are not, although, as Koser (2007) suggests, they have breached administration rules and regulations. It is difficult to assess the extent of irregular migration. People without legal status are likely to avoid completing census forms or engagement with governmental authorities. As a result, it is difficult to get an accurate picture of the number of irregular migrants. Estimates suggest that there were some 12 million irregular migrants in the United States in 2008 (Warren and Warren, 2013). In Chapter 5, we explore the politics of irregular migration in more detail. As we explore, irregular migration has created much political tension in the United States and other advanced economies.

It is important to point out that not all migration is permanent. There are temporary migrants, people who relocate to another country for a short time and then return to their country of origin. In more recent times, there has been an emphasis on the notion of circular migration, a process by which people migrate, return to their country of origin and then migrate again. With the increasing ease of transit, particularly for certain groups and for people with the proper documentation, circular migration is more possible now than in the distant past.

Throughout this book, we refer to different types of migrants and different types of migration that reflect the changing nature of the migration process. In many instances, we use broad categories to describe particular groups of migrants. For instance, we might refer to Chinese migrants, or Indian migrants, or Irish migrants. But it is important to remember that migrants come from particular communities. They might be from particular ethnic groups with different cultural nuances, different dialects. For instance, as Christiane Harzig and Dirk Hoerder (2009, 2) note, "[m]en and women of Chinese culture arriving in Canada between 1980 and 2000 hailed from 132 different countries prior to settlement and spoke some one hundred different languages and dialects." In developing an understanding of migration, we should also be cognizant that migrants are people with individual as well as cultural, social, and ethnic complexities. The immigrant and the emigrant are one and the same person with rich and complicated lives. In the next section, we explore some of the theories that explain why these people move.

WHY MIGRATION HAPPENS

Social scientists from various different disciplines and with different ideological and disciplinary perspectives have engaged in identifying what they consider to be the main reasons behind global migration (Brettell and Hollifield, 2007). People migrate for many reasons. Large numbers move to seek refuge from war and persecution. Most people migrate to reunite with family. Some move for work. In our examination here, we are mostly focused on this latter process. In this section, we examine a number of theories that offer an understanding of the migration process, with a focus on the movement of people because of economic reasons. First, we highlight work by one of the earliest theorists of migration, Ernest George Ravenstein, and then we integrate more contemporary scholarly work on recent understandings of global migration processes. For an excellent and highly extensive assessment of the different theories of migration, we advise the reader to examine Douglas Massey and colleagues' article, "Theories of International Migration: A Review and Appraisal," published in the journal *Population and Development Review* in 1993. We draw on this article as well as other sources to synthesize the major theories aimed at understanding the global migration process.

Ernest George Ravenstein was a geographer who, among other endeavors, published two influential papers on migration in the *Journal of the Royal Statistical Society*: one article in 1885, the other in 1889. As we explore in more detail in the next chapter, the rise of the industrial age in the mid-nineteenth century created large-scale changes in the working life of people, particularly in North America and Europe. Many people moved to industrializing cities for industrial jobs. Examining census data from 1871 to 1881, Ravenstein sought to identify and understand changing patterns of migration in the newly emerged industrial economy of the United Kingdom. As a result of his work, he identified seven laws of migration.

First, he found that most migrants move only a short distance, and second, if they move a long distance they tend to move to the largest urban centers of commerce. As we examine in Chapter 4, today there is large-scale movement of people to big cities that are central to the world economy. Third, Ravenstein found that migration flows tend to include a counter migration. In other words, migration includes inflow and outflow of people, so to understand the impact we should consider net migration. It is important to recognize not only the stock of migrants but also the flow of migrants when examining migration processes (Castles and Miller, 2009).

Fourth, migrants move to "absorption centers," leaving behind areas that are then filled by migrants from more remote regions. With this process, migration flows extend to, as Ravenstein (1885, 199) suggested, "the most remote corner of the kingdom." Today, there are few regions, countries, or communities in the world that have not been touched by global migration, either because people have left or entered. We aim to demonstrate this trend through our examination of different case studies of migration. Fifth, those areas that experience a net loss of people tend to lose these people to places that experienced a net gain in people. Sixth, people in rural areas are more migratory than residents of urban centers. This was true in Ravenstein's time and holds true today. Seventh, women tend to migrate more than men. In Chapter 3, we discuss in more detail the rise in the migration of women, and the relationship between gender and the migration process.

Ravenstein's laws not only created a lot of interest in his own time, but his work influenced later theories of migration, especially the theory that identifies so-called "push" and "pull" factors of global migration. "Push" factors coerce people to leave their place of origin, and "pull" factors entice people to a receiving country. "Push" factors include, for instance, low standards of living and a lack of job opportunities in the sending country. "Pull" factors include, for instance, the demand for labor, good employment opportunities, high wages, and political freedom in the receiving country.

Such "push" and "pull" factors are part of a broader understanding of migration that is embedded in the neoclassical economic tradition. Neoclassical economics considers both micro and macro theories of migration (Massey *et al.*, 1993). Macro-economic theory emphasizes the role of wage differentials between countries as the primary explanation for migration (Solimano, 2010). Countries where there is an oversupply of labor relative to capital have low wages; while countries with less labor than needed have high wages. Neoclassical macro-economic theory suggests that these wage differentials will cause workers to migrate from low-wage to high-wage countries. Because of this shift in the labor force, countries with a previous oversupply of labor will see wages rise as this supply is reduced with the out-migration of workers to high-wage countries. Similarly, the introduction of new migrants into countries with an undersupply of workers will cause wages to decrease in those countries. Migration will cease once it leads to an eventual state of international wage equilibrium.

At the micro-economic level, individual workers make decisions based on their assessment of the costs and benefits of migration. One of the major benefits of migration for the individual is higher wages in the destination country. However, there are costs associated with this move. For instance, there are the material costs of traveling. There are also the psychological costs of possibly leaving behind family and friends. Neoclassical micro-economic theory suggests that people will move if the expected rate of return from the higher wages and employment opportunity in the receiving country is greater than the costs of migration. George Borjas has written extensively on the economics of migration. As he explains (1989, 461):

> Neoclassical theory assumes that individuals maximize utility: individuals "search" for the country of residence that maximizes their well-being. ... The search is constrained by the individual's financial resources, by ... immigration regulations. ... In a sense, competing host countries make "migration offers" from which individuals compare and choose. The information gathered in this marketplace leads many individuals to conclude that it is "profitable" to remain in their birthplace. ... Conversely, other individuals conclude that they are better off in some other country. The immigration market non-randomly sorts these individuals across host countries.

The neoclassical economic model of migration is very influential, and there are a plethora of studies whose empirical analyses are based on an understanding of migration as a rational choice where

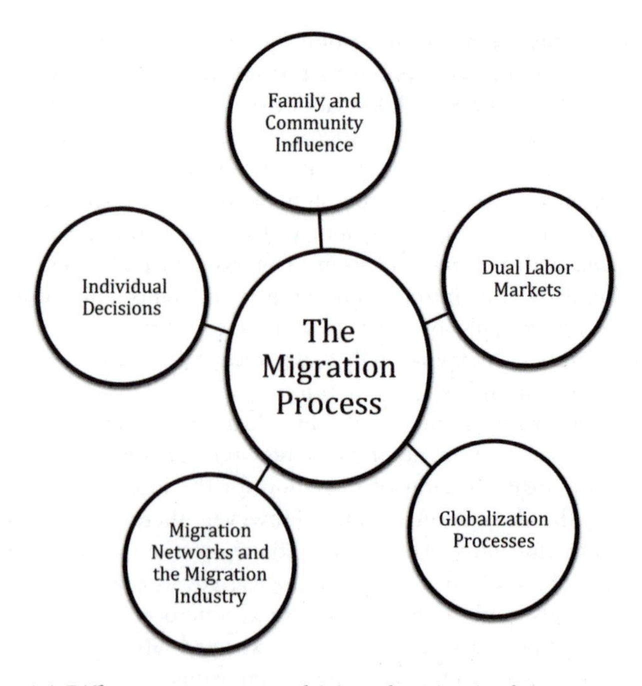

Figure 1.1 Different components explaining why migration happens.

individuals weigh the costs and benefits of relocating. The neoclassical model has also been influential in the realm of public policy. Many times immigration policy rests on government manipulation of the labor market. For example, governments will develop policies that encourage the in-migration of workers with specific skills if there is a perceived undersupply of workers with those particular skills (Massey *et al.*, 1993).

In Figure 1.1, we illustrate the major components of the migration process. The neoclassical view emphasizes the component we label "individual decisions." There is some criticism of the neoclassic model as an explanation of migration. Stephen Castles and Mark Miller (2009, 23) suggest the following:

> It seems absurd to treat migrants as individual market-players who have full information on their options and freedom to make rational choices.

Instead migrants have limited and often contradictory information and are subject to a range of constraints (especially lack of power in the face of employers and government).

It is important to recognize that while many poor people want to escape their low standard of living, not all poor people migrate. In fact, the poorest people generally do not migrate. As Peter Stalker (2001, 20) suggests, "Côte d'Ivoire, for example, with a per capita GDP (Gross Domestic Product) of $1,600 is by global standards a poor country yet relatively few of its people leave." There can be a variety of dynamics that influence why a person migrates.

This brings us to a second component, "family and community influence." This component is part of a broader discussion of the new economics of migration. This recent approach to an understanding of the migration process focuses less on individual decision-making, but rather it suggests that families or even communities make migration choices together. Individuals within a family or community migrate because the family or community decides it is best for that collective unit. Here the role of remittances is important. Migrants working in foreign countries send money back home to their families and communities.

The new economics of migration stresses the importance of remittances in offsetting, for instance, income insecurity associated with the lack of insurance systems in less developed countries (Massey *et al.*, 1993). Because many developing countries do not have crop insurance or unemployment insurance, if crops fail or someone in a family or community loses their job, the results can be devastating. Money sent home by migrants offers protection from such income instability among migrant families and communities. In addition, these countries of origin often lack sophisticated credit systems to enable capital investment in farming or other economic activities. According to the new economics of migration, the money sent back home by migrants adds to the collective income of vulnerable communities and can be used to invest in local economies. Together the family or social group decides one of their members should migrate to diversify their sources of income and capital. In more recent years, there has been an emphasis on the role of migration in development within less developed countries. We examine this relationship in Chapter 4 when we discuss migrants and their role in the global economy.

Both the neoclassical economists and theorists supporting the new economics of migration focus on the importance of micro-based decisions for understanding global migration. For the neoclassical economists, decisions are made by the individual migrant; for supporters of the new economics of migration, the decision-making unit is the family, household, or community. In both cases, the decision-making process aims to maximize income and benefits and minimize risks and costs associated with migration. The focus is on agents acting to make their own choices, albeit within an environment with certain constraints or particular characteristics.

In his influential book, *Birds of Passage: Migrant Labor and Industrial Societies*, Michael Piore (1979) provides an alternative view to the neoclassical economists' understanding of global migration by moving beyond the behavior of migrants to suggest that larger social, economic, and political structures of advanced economies are central to the global migration process. In his book, he suggests that rather than migrants making the decision to move to another country, private industries actively recruit foreign workers. In particular, they recruit migrants to do work that native populations are unwilling to do. Piore (1979) suggests the reasons why employers recruit migrants are inherent in the economic structure of advanced economies. The labor market in these economies is segmented into, on the one hand, capital-intensive primary-sector jobs with skilled workers and, on the other hand, a more labor-intensive secondary sector where jobs are unskilled, poorly paid, and unstable. It is difficult to attract native workers to the low-skilled, low-wage positions. They are drawn to the primary-sector positions that are better paid and more stable. To make up for the shortfall within the labor-intensive secondary job market, employers actively recruit migrants. In Figure 1.1, we identify this process as "dual labor markets."

According to the dual labor market theory of migration, global labor migration is largely driven by the demand for low-skilled workers. In contrast to the classical economics and theories of the new economics of migration, dual labor market theory sees the bifurcation of the labor market into low- and high-skilled labor as central to the migration process. As such, according to this theory, it is important to consider the role of employers and principal labor market structures.

There are certainly contemporary and historic examples of active recruitment of migrants for low-skilled jobs by employers. Railroad companies and contractors, for instance, recruited Irish migrants during the period of railroad expansion in the mid-nineteenth century in the United States. Today, the U.S. meatpacking industry seeks out migrant workers from Mexico, El Salvador, Guatemala, Peru, and other Latin American countries for dangerous, low-paying jobs. But it is not just employers who recruit migrant low-skilled workers. In 1942, the first group of workers from the Bracero program arrived in the United States from Mexico. They were actively recruited during World War II for agricultural labor. Ending in 1964, an estimated 4.5 million Mexicans entered the U.S. through this program and established networks enabling more migrants to follow. More recently, Canada, through its Seasonal Agricultural Workers Program (SAWP), has recruited temporary workers from Mexico for low-skilled agricultural and manufacturing jobs. However, it is not only low-skilled workers that are recruited by employers and governments. There is active recruitment of highly skilled technicians. For example, in the United Kingdom and other advanced countries, healthcare workers are sought. The needs of the global market economy are varied.

This brings us to the fourth component we label "globalization processes." In his multi-volume composition *The Modern-World System*, Immanuel Wallerstein (1974) traces the history and development of the modern capitalist world economy. He suggests that modern capitalism grew slowly over time, beginning in the sixteenth century. At the heart of his research is the idea that, with capitalism's expansion, less developed "periphery" regions of the world become incorporated into a global capitalist economic system dominated by more developed "core" regions. *World systems theory* emphasizes the notion of capitalist expansion from the developed "core" into the less developed "periphery" of the globe. In this vein, the penetration of capitalism into less developed regions creates turmoil for those regions and inevitably leads to a more mobile population who are thus prone to migration.

Douglas Massey and his coauthors (1993) examine in more detail the different ways in which disruption takes place in less developed regions. For example, farming practices change dramatically with the introduction of capitalist agricultural production in the form of cash crops. Large-scale industrial practices and land consolidation undermine traditional systems of land tenure, and increasing agricultural mechanization creates unemployment among farm laborers. These processes contribute to a more mobile labor force among small agrarian communities. This is just one example of how the expansion of global capitalism into less developed regions of the world creates a vulnerable population prone to migration.

Beginning after World War II and right up until the 1970s, there was great interest in encouraging economic development among nations regarded as not sufficiently developed. Less developed countries of the South were encouraged to engage in import-substitution. Economic policy stressed that Southern nations develop their own industries so they would no longer have to rely on imports from the developed world. Import-substitution had the adverse effect of increasing foreign direct investment by the developed world in the newly emerging industries of less developed countries. Firms from developed countries reaped large profits from these investments. Advocates of dependency theory suggest that development policies such as import-substitution industrialization resulted in an increase in dependence by less developed nations of the global South on more developed nations of the global North.

Dependency theory and world systems theory have been very influential, especially in the study of more recent globalization processes. Globalization theory emphasizes the interconnectedness and cross-national flows of capital finance, trade, and, in the context of migration, people. Central to the emergence of these global processes is economic integration. Multinational corporations (MNCs) are of particular importance. They operate in different countries, and, in many cases, they shift their operations to less regulated zones. With recent globalization, markets are sought and opened to worldwide competition where the emphasis is on privatization and a free market economy. Integration of the global economy is guided by liberalization and deregulation. The result is further penetration of capitalism across the globe (Petras and Veltmeyer, 2001).

These globalization processes have emerged in the context of neoliberalism (Castles and Miller, 2009). David Harvey (2007) discusses neoliberalism as a combination of old-style liberalism's commitment to individual liberty with the neoclassical economic emphasis on primacy of the free market and opposition to any state intervention in the economy. The intellectuals behind the promotion of neoliberalism include economists such as Friedrich von Hayek, Ludwig von Mises, and, later, Milton Friedman. The economic doctrine of neoliberalism places great faith in the free market. Free trade and a strong commitment to deregulation to eliminate constraints on trade and the market are central tenets of this doctrine. For those who advocate neoliberalism, the state is to be subordinate to the economy. Taxes should be low, especially for businesses and industry, and welfare spending should be minimized.

International institutions such as the International Monetary Fund (IMF), the World Bank, and the World Trade Organization (WTO) have been instrumental in the promotion of neoliberalism across the globe. Nations that have received aid from these organizations have had to restructure their economies according to the neoliberal doctrine. Part of a program of structural adjustment, weaker nations and societies are required to limit social welfare and open up their economies to what Castles and Miller (2009, 53) refer to as "the cold winds of competition." Today, we see the neoliberal agenda promoted in the so-called austerity measures introduced across European economies devastated by the most recent recession.

Proponents of neoliberalism hold that limiting restraints on the market will lead to economic growth in poor nations. There will be a trickle-down effect that will reduce poverty and eventually result in poorer nations becoming wealthier nations. The reality, however, is that global inequality is on the rise. As Branko Milanovic (2007, 39), lead economist with the World Bank, suggests, global inequality is "probably the highest it has ever been." One of the major concerns regarding globalization in the neoliberal era is that it leads to profound social inequality within and between countries. With neoliberalism, labor has become cheap and labor markets much more precarious. China and the former Soviet Union have now embraced the global capitalist economy, and the workforce has expanded as a result. Large corporations have benefited from a cheap, culturally divided and more flexible labor force. The reality is a deepening of uneven development across the globe.

In Chapter 2, we discuss the impacts of recent globalization on the patterns of migration, but a major driving force is the growing inequality between countries. Often it is the growing inequality between advanced economies of the North and the rest of the globe, referred to as the North–South inequality (Castles and Miller, 2009) that is highlighted. As we indicated earlier, the reality is a little more nuanced. The new industrial economies of Asian and Latin American nations—such as China and Brazil—are expanding, and within the global South, new wealthy elites have emerged, while in the industrial parts of the global North workers are struggling. There is also a great divide between rural and urban incomes in industrializing nations such as India and China. Inequality of different forms is leading to increased mobility as people from poorer regions, economic sectors, and nations move to growth areas within a region or to the more advanced economies, particularly in the North. This movement is especially necessary in a neoliberal era when social welfare programs are lacking and labor unions have lost their power.

As we emphasize the role of globalization in migration, it is also important to note that it encourages the free movement of capital and services, including the global dispersion of information. This makes it easier for migrants to find out about different countries before they move. Migrants have the ability to develop networks using new technologies such as the Internet and mobile phones, and the transition to a new country can be made more comfortable with the use of Skype, e-mail, and other types of global communication. Modern technologies make it easier for people to develop networks. This brings us to the fifth and last component in Figure 1.1: "migration networks and the migration industry."

Douglas Massey and his colleagues (1993, 448) define migration networks as "sets of interpersonal ties that connect migrants, former migrants, and non-migrants in origin and destination areas through ties of kinship, friendship and shared community origin." Migration network theory stresses the importance of family and community ties to the flow of migrants to particular areas. Networks facilitate the process of migration settlement and the formation of immigrant communities. These networks are a form of social capital. Migrants can draw on the ties with family and community members who migrated before them for employment opportunities. As the migrant network develops within a particular community, migration becomes, as Massey and his colleagues suggest (1993), self-perpetuating. Each new migrant builds the social network for other migrants.

Early scholars of migration networks coined the term "chain migration" (MacDonald and MacDonald, 1964). Drawing on the example of large-scale migration from Italy to the United States in the early twentieth century, John MacDonald and Leatrice MacDonald (1964, 82) defined chain migration as "that movement in which prospective migrants learn of opportunities, are provided with transportation, and have initial accommodation and employment arranged by means of primary social relationships with previous migrants." Migration network theory stresses individual and household decision-making processes but also recognizes that each act of migration to a particular community alters the context in ways that encourage future migration to that community. For Massey and his colleagues (1993, 461) there is a process of cumulative causation where "migration sustains itself in such a way that migration tends to create more migration."

More recent research related to the building of migration networks focuses on the role of political and economic institutions in the migratory process. A migration industry has emerged to guide and enhance migration. This industry is made up of lawyers, recruitment agencies, smugglers, non-profit organizations, entrepreneurial agents, and so on. Some of the members of this migration industry can exploit vulnerable migrants, especially those that enter the country without proper legal documentation. Some migrants have found themselves indebted to recruiters, or they have lost their life savings as swindlers leave them stranded in a new country without the promised job or resources. Others can help migrants adjust to their new life and help them gain legal status within their new destination country. As Castles and Miller (2009) suggest, migration agents include members of the migrant community, including shopkeepers, teachers, community leaders, and others who help migrants settle in their new society.

In summary, the processes behind global migration are multifaceted and complex. For neoclassical economists and those who stress the new economics of migration, decision-making by individual or community agents is central to migration. A person or community decides to migrate or not based on an assessment of the costs and benefits of migration. For proponents of dual labor market theory, world systems theory, and globalization theory, the most important reasons for global migration relate more specifically to increasingly global economic and labor market structures, and inequality between regions, nations, and economic sectors. For others, institutions and the migration industry guide the global migration process.

Douglas Massey and his colleagues (1993) take the view that there is likely no one theory that explains global migration. We tend to agree. As Massey and his colleagues (1998, 455) state, "rather than adopting the narrow argument of theoretical exclusivity, we adopt the broader position that causal processes relevant to international migration might operate on multiple levels simultaneously, and that sorting out which of the explanations are useful is an empirical and not only a logical task." And in many respects, there is much to be done empirically. To understand the reasons why migration occurs in the ways that it does, further empirical investigation is required and an approach that considers the varied facets of this complex process is needed. It is also important to keep in mind that the current global capitalist mode of production creates uneven development across the globe, and that the inequality between nations creates structural imbalances that lead to a relocation of people from poor to rich countries.

PLAN FOR THE BOOK

Our main goal in this book is to provide the reader with a basic understanding of the key elements of global migration from different disciplinary perspectives. To achieve this goal, we examine past and recent global geographic patterns of migration; the impact of migration on receiving societies; the role of migration in the global economy; and the public policy and political responses to migration. We intersperse the text with case studies to illustrate the various aspects of global migration.

In Chapter 2 we examine the history and geography of migration. More specifically, we consider how the geographic patterns of global migration have changed over time. We examine patterns of migration from the colonial period of European Imperialism through to the current era of global integration where migration has touched many different countries in many different regions of the world. Nations that may have previously experienced emigration now receive immigrants from other nations. Ireland is an informative example. A nation with a long history of emigration to nations such as the United Kingdom, the United States, Germany, and Australia, Ireland began to experience the inflow of people from countries such as Poland, Nigeria, Romania, China, and Pakistan during the economic boom period of the 1990s. While in-migration to Ireland has slowed and emigration from Ireland has re-emerged, the government and society of Ireland has been greatly impacted by a new foreign-born population living inside its borders. Migration alters with different times and different economic and political changes.

In Chapter 3, we examine the impact of migration on the destination society. When migrants settle in a new place, especially in relatively large numbers or relatively quickly, they dramatically transform that place. In some cases, with acceptance of the new cultural diversity, migrants will form ethnic communities that are celebrated by the new multicultural society (Castles and Miller, 2009). In other cases, new migrants may be rejected by the new society as a threat, in particular to national security and national identity. In more recent times, migrants, especially irregular migrants, have been perceived as a criminalized threat to national security. Since the attacks of September 11, 2001 in New York, the bombings in London in 2005, those in Madrid in 2004, and then more recently the bombing at the Boston Marathon in 2013, there has been increasing concern that migrants are a threat to national security. These terrorist attacks have had a tremendous impact on societal perceptions of migrants. With increasing militarization and securitization more generally, the attitude and laws surrounding migration have changed. In response, many migrant groups have felt increasingly threatened and excluded from their host countries. This chapter explores these perceptions, concerns, and consequences in a post-September 11 world.

As part of this discussion in Chapter 3, we focus on processes related to migration, including assimilation, ethnicity, segregation, multiculturalism, and transnational migration. We examine the intersection between gender and migration, focusing on recent discussions around the feminization of migration and the particular concerns of female migrants. Finally, we explore the nexus between migration and sexuality, focusing in particular on the policy environment for lesbian, gay, bisexual, and transgendered (LGBT) individuals.

In Chapter 4, we examine the relationships between migrants and the global economy. We explore the relationship between migrants and the labor market, particularly their effect on the wages and employment of native workers. We also discuss the ethnic economy, focusing in particular on an examination of ethnic niches and ethnic enclaves defined in this chapter. One of the many celebrated aspects of globalization is the free flow of capital and information across nations. Yet behind this movement are people who migrate in search of employment. In particular, of importance to migration processes and patterns are the large global or gateway cities with strong links to the expanding global network of commerce and finance. Cities such as New York, London, Los Angeles, and Sydney are long-established immigrant gateways. They attract people from all over the world because of their importance as centers of global trade and finance. Other cities such as Dublin or Washington, DC can be categorized as emerging immigrant gateways (Price and Benton-Short, 2008). Such new nodes in the global economy have recently attracted people from other countries. We focus in particular on the nature and effects of the movement of high-skilled migrants. Finally, in this chapter, we explore the nexus between development and migration, particularly in the context of remittances.

In Chapter 5, we focus on the political and policy implications and responses to global migration. We provide the reader with a broad historical view of migration control. We examine the nature of irregular migration, and the growing number of refugees and asylum seekers globally. As part of the discussion about globalization, there is an emphasis on flows, in particular the flow of people, things, and information across national borders. However, just as there is an increase in movement and a growing economic linkage at the global level, there are also structures that impede or block flows, particularly flows of people (Ritzer, 2011). An example is the border preventing the inflow of irregular migrants from Mexico into the United States. We focus on more recent concerns, particularly in the developed world, over issues of border security in the wake of the so-called War on Terror. Migration policy has generally toughened throughout the world. In this chapter, we examine reasons for migration control and its impacts.

In Chapter 6, we summarize the major aspects of migration covered in the book, and then we consider the important influences and elements of future global migration. There are concerns about climate change and how this might impact the movement of people in the future. Our global world is more connected digitally than ever before. There is a mobile elite living in different places at different times in their lives for purposes of a new lifestyle. This chapter examines the future tendency for more environmental migrants and lifestyle migrants. In an increasingly global world, migration in its different forms is destined to continue to rise. Within this context, we also discuss the continuation of irregular migration.

In this book we emphasize certain aspects of global migration, and focus on certain geographic patterns and places where migration has occurred. Of course, we cover some topics in more depth than others. We explore some geographic areas in more detail than others. Trying to examine every aspect of the vast and fascinating topic of global migration is an enormous challenge. We have made a good faith effort and offer suggestions for further reading on aspects we may have missed or have not covered in as much depth as we would have liked. We hope you enjoy the results.

GUIDE TO FURTHER READING

For migration research purposes, see the website from the International Organization on Migration (IOM): http://www.iom.int/cms/en/sites/iom/home/what-we-do/migration-policy-and-research/migration-research-1/migration-profiles.html (accessed November 14, 2013).

For more detail on Marxist theories of migration, see Vogel, R. (2013). "Marxist theories of migration." In *The Encyclopedia of Human Global Migration* edited by Immanuel Ness and Peter Bellwood, Hoboken, NJ: Wiley-Blackwell Publishers.

For information on migration definitions, see Khalid Koser (2007). *International Migration: A Very Short Introduction*. Oxford, UK: Oxford University Press, Chapter Two.

MIGRATION ACROSS THE GLOBE

Global migration has a long history that dates back to the beginning of humankind. In descriptions of major historic developments, scholars have focused on a number of key periods of time and, in many cases, they have tied migration during these periods to large economic and political forces (Massey, 2003; Koser, 2007). In this chapter, we examine migration corresponding with a timeline from 1600 onward, as colonial Europe expanded. We also explore patterns of migration during the modern era of large-scale industrialization, and we end with an examination of patterns of migration during the current age of postindustrial globalization. Our goal in this chapter is to identify the larger geographic patterns of migration over the past few centuries.

MIGRATION DURING EARLY COLONIAL EXPANSION

The period of European colonial expansion originated in the seventeenth century and continued into the middle of the nineteenth century, when several countries, including Spain, the Netherlands, Portugal, Britain, and France, established colonies in the continents of Asia, Africa, the Americas, and Oceania. During this period, there were three forms of migration: the migration of free people; the forced migration of slaves; and finally, the migration of indentured servants. The indentured servant of the colonial period was bound by very strict labor contracts, and was migrated by force, in many cases, from their country of origin to work under very poor conditions with no pay until such time as their contract expired. Slavery, indentured servitude, and free migration were essential to the rising economic power and success of countries such as Britain, Spain, Portugal, the Netherlands, and France, and later the former colonies of the United States and Australia. Those people who migrated freely before the early nineteenth century were mostly those of economic means; they were wealthy people who could pay the high transportation costs necessary to arrive safely in the New World.

The colonies were, from the outset, great sources of raw materials for commodity production in Europe, specifically with a wave of rising consumer markets there (Blackburn, 2010). Plantations and mines developed across countries of the Caribbean, South America, North America, and parts of Asia and Africa. Given preindustrial technology, these plantations and mines required a large amount of cheap and free labor to be productive. The production of tobacco, sugar, coffee, and cotton relied heavily on the slave trade and, in the case of indentured labor, the recruitment of large numbers of workers bound to strict labor conditions.

According to Barry Chiswick and Timothy Hatton (2003), an estimated 700,000 Europeans migrated to North America and the Caribbean between 1650 and 1780 and, of these, between one-half and one-third were indentured servants. Indentured servitude was an important element of migration during the early colonial period. Akin to this form of forced migration was the relocation of convicts from Britain and Ireland to Australia. Some 80,000 convicts arrived in New South Wales between 1788 and 1840.

By the end of the eighteenth century, indentured servitude in the Americas slowed as the slave trade expanded. The slave trade included the forced migration of people from Africa to the Americas. Between the sixteenth and nineteenth centuries, approximately 12 million African slaves were imported to the Americas (Rediker, 2008). This resulted in long-lasting effects on the transformation of colonies in the Caribbean, United States, and South America. With imperialist expansion, ships sailed with manufactured goods from ports in England and France to West Africa.

These manufactured goods were delivered and replaced with African slaves who had been abducted by force or purchased, in return for the manufactured products, from tribal chiefs or slave traders. These same ships again set sail, bound for the Americas where the slaves were sold for cash, which was then used to purchase raw materials and products from the plantations. The ships made their return trip to Europe to complete the triangular trade system. The slave trade that evolved during the seventeenth and eighteenth centuries was heavily tied to the rising global network of merchant economies and the construction of a world market.

The trafficking of slaves was legally abolished within the British Empire in 1807, and by 1815, other European countries had followed suit. Slavery itself did not finally end in the British colonies until the 1830s, and then some 30 years later in the Dutch colonies. After a bloody Civil War in the United States, the institution of slavery was abolished in the southern states in 1865. It was not until 1888 that slavery was abolished in Brazil—the last nation in the West to do so. The reasons for abolition are complex. Historian Robin Blackburn (2011) suggests humanitarian concern, but he also posits that, more importantly, political movements toward national independence in the colonies combined with revolt and resistance by the slaves themselves led to the eradication of this horrific institution. However, it is important to recognize that forms of slavery and indentured servitude still exist today. We refer to this in Chapter 5.

During the period of colonial expansion, with the legal abolition of slavery in the colonies, there was resurgence in indentured servitude (Northrop, 1995). Beginning in the latter half of the nineteenth century, indentured workers began to replace slaves as important sources of cheap labor in the plantations and mines in colonial nations. But rather than extracting people from Europe, as in earlier periods, these indentured migrants were drawn from places such as China, India, Africa, parts of the Pacific Islands, Caribbean, and North and South America. Probably the most significant was the recruitment of Indian and Chinese people. The destinations for indentured migrants from China included countries such as Hawaii, Peru, Brazil, and the U.S. (Shimpo, 1995). As for Indian indentured migrants, some ended up in the countries of Mauritius and Fiji (Thiara, 1995). The British also recruited workers from India to work on the sugar plantations in countries of the Caribbean as part of the so-called "coolie system." The overwhelming majority of indentured servants from India did not return to India once their contracts expired. They disliked the idea of returning to the lowly positions they held in the traditional caste system. They remained in the Caribbean in the hope of acquiring land and new opportunities. As Castles and Miller (2009) point out, indenture reflects the classic "divide and conquer" strategy of the colonial powers. In the case of the Caribbean, for instance, indentured servants from places such as India were positioned against the African population that had arrived as slaves decades earlier. To this day, skin color has deep implications for identity among Caribbean groups, with the darkest-skinned populations experiencing most discrimination. The use of contract labor lasted, in some cases, until the early twentieth century (Engerman, 1986). Upon completion of their contracts, many indentured servants remained and settled in their new country.

There were three types of migrant in the colonial period: free people, indentured servants, and, of course, slaves. During the 1820s, the numbers of slaves that were brought to the Americas each year far outnumbered the free migrants. However, by the 1840s, the in-movement of free migrants to the Americas greatly exceeded that of slaves. Many of the free migrants that entered in the early part of the nineteenth century were artisans, laborers, farmers, craftsmen, and former sailors. Prior to 1820, steerage costs were so great that only the more advantaged could afford to make the voyage. By the late 1840s and early 1850s, many more unskilled laborers migrated to the Americas, and particularly in the 1840s, a large contingency of unskilled and very poor Irish migrants migrated as a result of the Great Famine in Ireland. As we discuss in the next section, this shift to free migration was greatly influenced by the rise of industry and technological changes in the Americas and Europe.

In short, migration during the early colonial era was characterized by a large outflow of population from Europe to parts of the colonized world during the period of colonial expansion from the sixteenth to the beginning of the nineteenth centuries. Slaves from Africa were forced to migrate to the Americas. Indentured servants from Europe and, in later periods, Asia were recruited in large numbers. Many migrant workers who moved either by force or voluntarily to the colonies died during the long journey or from warfare and tropical illnesses. Those that survived went on to have a tremendous impact on the culture, economy, and development of both Europe and its colonies. The early European emigrants settled with devastating effects in the indigenous populations of colonized countries. As part of colonial domination, native populations of the Americas, Oceania, Asia, and Africa were exploited, and, in many instances, large-scale genocide resulted in the death and destruction of indigenous peoples and their cultures.

MIGRATION DURING INDUSTRIAL EXPANSION

The accumulated capital reaped from colonization was used to further expand and develop manufacturing industry and commercial agricultural enterprises both in Europe and the former colonies, especially North America. During this period, the rise of a global economy centered on industrial production in Europe, North America, and Japan, as well as a reliance on the export-rich periphery areas of Southeast Asia and the South Pacific. Migration patterns were heavily influenced by the integration of industrial economies within Europe, the former colonies, and other parts of the world.

In this section, we examine three large movements of people from about the mid-nineteenth century through the 1950s. First, we discuss a continued transatlantic migration of Europeans to the Americas. Second, we examine the nature of internal migration within Europe. Third, we consider briefly migration in periphery countries, particularly to Southeast Asia.

EUROPEAN MIGRATION TO THE AMERICAS

The transatlantic pattern of migration from Europe to North America, beginning during the colonial period but intensifying in the mid-nineteenth century, is one of the more renowned movements of people in human history. Between 55 and 58 million people migrated from Europe to the Americas between 1846 and 1940 (McKeown, 2004). Almost two-thirds of these migrants settled in the United States, and the remainder arrived in Canada, Argentina, Brazil, and Cuba.

In the period between 1880 and 1900 alone, 1.6 million European migrants arrived in Brazil; half of these people were from Italy and a quarter from Portugal. Then, between 1908 and 1936, some 1.2 million migrants from Italy, Portugal, Spain, and Japan arrived by ship in the port of Santos, Brazil. Many of the initial migrants to Brazil worked in the coffee plantations (Klein, 1995). Many European migrants returned to their home countries after a period of time, staying only temporarily. During the early 1900s, Brazil became the main migration destination for the Japanese. The Japanese government offered financial support and organized migration for many of its citizens during this period. These Japanese, like the Italians previously, worked initially in agriculture. Between 1910 and 1929, there were close to 86,000 Japanese immigrants in Brazil.

There were several boom periods of in-migration to Argentina, especially between 1880 and 1929. During the years 1905 and 1914, close to 3 million migrants arrived. Between 1920 and 1929, there were an additional 3 million immigrants (Adelman, 1995). Many of these migrants came from Italy and Spain. The fertile area of the pampas contributed to Argentina's strong agricultural economy, and many early migrants worked in this sector. Some early migrants did not remain permanently in Argentina, but rather they migrated for seasonal work in agriculture, returning to their home countries after the harvest. The slow pace of industrialization in Argentina was a contributing factor to return migration.

Migration to parts of South America is a smaller proportion of the transatlantic migration than the large-scale movement of Europeans to the United States. As Robin Cohen (1995, 77) suggests, the great migratory shift to the U.S. is the "stuff of legend." Images of people entering New York's Ellis Island from ships are etched into thousands of history books (Figure 2.1). The rags-to-riches immigrant story has its roots in this period and still offers a powerful expression of America's identity as a nation today. The number of immigrants to the United States began to soar after 1850, peaking in the 1920s. The annual average of migrant passengers entering the United States was 12,847 in the 1820s; 53,100 in the 1830s; 152,760 in the 1840s; and 275,458 in the 1850s. These four decades signaled a tremendous growth in the number of immigrants to the United States. Of those migrants who reported

Figure 2.1 Immigrants just arrived from foreign countries—Immigrant Building, Ellis Island, New York Harbor.
Source: Library of Congress Prints and Photographs Division Washington, D.C. 20540, USA.

their country of origin, by the end of the 1850s about 96 percent were European, with the remainder from Mexico, the West Indies, and South America (Hatton and Williamson, 2008). After the depression of the 1890s, migration to the U.S. once again soared. In the first decade of the 1900s, 9 million immigrants entered the country.

Transatlantic migration was heavily influenced by changes in the global economy of the time. With early industrial expansion, there emerged en masse a new class of wage laborers bound for the factories that arose in cities such as Manchester and Glasgow in the United Kingdom, Pittsburgh and Baltimore in the United States, to name but a few. In particular, early industrial growth in Britain was spurred by the availability of a labor class of displaced farmers or farm workers forced from rural to urban areas because of the consolidation of small farms into large agricultural estates. The labor classes included displaced artisans and craftsmen outdone by competition from larger-scale manufacturing. The emergent capitalist mode of production caused great disruption and unsettled large numbers of people who moved either within their own country or abroad. This was the time when geographer Ernest George Ravenstein conducted his classic study of migration in the United Kingdom. Ironically, during the height of Britain's industrialization, many of the displaced labor classes migrated to North America (Castles and Miller, 2009).

In the middle of the nineteenth century, those people that made the voyage to the Americas included unskilled labor classes. These migrants were not necessarily the poorest of the poor, with the exception of the Irish. Between 1845 and about 1853—the period of the Great Famine in Ireland—about 1.5 million Irish left their homeland because of hunger and the potato blight, many leaving on "coffin ships" to America's Eastern Seaboard. As Robert Scally describes (1995, 81):

> It is the air of panic that has given the famine emigration its distinct character. The great majority consisted of a new sort of emigrant, one who had been seen boarding the ships before, but never in such numbers and such misery. And unlike the victims of any previous famine in Europe, these refugees were passing in the millions through the core of a great empire and the most dynamic centers of the world economy at the time ...

Even given the tremendous wealth generated by British colonialism and industrial expansion, there was desperation among many of the Irish arriving on the shores of the United States during the mid- and late nineteenth century.

Migration to the Americas during the period was greatly impacted by many technological innovations. Advances in steamship technology, for instance, assured migrants from Europe of a fast and safe ocean crossing. Beginning in the 1870s, steamships became the primary means of transportation for European migration to the United States. Railroad expansion was also important. The railroads played a vital role in the development of commodity markets throughout North America. In 1862, Abraham Lincoln signed the Pacific Railway Act authorizing the federal government to fund railroad expansion westward across the United States. Subsidised for each mile of railroad track, the railroad companies sought to lay down as many tracks as they could, as quickly as possible. They recruited Chinese, Irish, and Mexican migrants to do the work, and, of course, railroad expansion to the west meant that more immigrants could move well beyond the ports of the Eastern Seaboard. Railroad development in Europe made ports more accessible to migrants wishing to take a steamship to the Americas. So, by the midpoint of the nineteenth century, the cost of long-distance journeys across the Atlantic Ocean and within North America declined significantly (Hatton and Williamson, 2008).

The railroad opened up the western parts of the United States. Scandinavians and Germans were the first groups to move westward, settling in Midwestern states. Initially, these migrants relocated to the Midwest because there was cheap land available to farm, but later they worked in factories of growing Midwest cities such as Milwaukee, St. Louis, and Cincinnati. After the 1880s, migrants from southern and eastern Europe began to arrive on American shores. By 1910, these groups of Europeans constituted about three-quarters of all migrants in the USA. People from Poland, Hungary, and Italy worked in coal mines and steel mills while Russian and Polish Jews worked in the textile industries in New York City.

As the industrialization of North America proceeded during the late nineteenth and early twentieth centuries, most migrants worked in factories rather than on farms, and they gathered in cities rather than in rural areas. According to Christiane Harzig and Dirk Hoerder (2009), until 1850, one-third of people entering the U.S. migrated to farms; after 1890, almost all migrants were engaged in industrial work. Between 1880 and 1920, employment in the U.S. manufacturing sector expanded from 2.5 to 10 million workers. During this same period, the number of immigrants to the U.S. increased from about 7 million to close to 14 million. Industrial expansion demanded many unskilled laborers. The work of these laborers led to ever-more industrial expansion and the need for more laborers.

Migration to the U.S. slowed significantly in the 1920s. The Immigration Act of 1924 limited the annual number of immigrants who could be admitted from any country to 2 percent of the number of people from that country who were already living in the U.S. in 1890. The main aim of the legislation was to restrict the number of migrants mainly from southern Europe. These restrictive migration policies in the U.S. impacted the extent and nat migration to Brazil and Argentina that we examined earlier.

Discrimination against various immigrant groups has often be an element of public policies on migration, as we will explore in Chapter 5. Yet, immigrants helped build the economy of the U.S. As Robin Cohen (1995, 79) states, "[they] provided the essential grist to the American mill just when the country was on the verge of becoming a 'great power'." Migration to the U.S. was certainly influenced by the changing economy and political climate in Europe, as well as the growing needs of a newly forming industrial economy. From 1920 until the mid-1960s, the policy environment restricted migration to the U.S. Annually, between 1951 and 1960, an average of 250,000 people entered the country, which was a sharp decline from the 880,000 immigrants that entered annually from 1901 to 1910. As we explore later in the chapter, migration to the United States increased again with changes in public policy and the global economy.

EUROPE'S INTERNAL MIGRATION

While transatlantic migration accounts for much of the movement of Europeans between 1840 and 1940, there were also various movements within Europe. As the earliest industrialized country in Europe, Britain was one of the first European countries to experience migration during this period. Factory towns sprung up around Britain beginning in the early nineteenth century, absorbing a large number of international migrants from two sources: the Irish peasantry, and Jews who arrived as refugees after the pogroms of Russia between 1875 and 1914 (Castles and Miller, 2009).

The Irish famine of 1846 and 1847 generated mass emigration to the United States. It further caused a large-scale movement of Irish people to Britain. According to Castles and Miller (2009), some 700,000 Irish were living in Britain by 1851, which accounted for approximately 3 percent of the population of England and Wales and about 7 percent of the population of Scotland. The Irish migrants in Britain were primarily peasant stock, and they worked in the large factories and in the construction sector of the industrial cities of Britain. Cities such as Liverpool, Glasgow, Manchester, London, and Birmingham were primary destinations. They also re of

n

anals and railways, gaining the moniker "nav-
from the word "navigator" (Castles and Miller,
period, the Irish suffered from much dis-
were often the target of hostility among the

e Jewish refugee population from Russia settled lar-
n London, more specifically in London's East End. They worked mostly in the clothing industry. The forced migration of Jews from eastern Europe raised concern among the British, so much so that a racist backlash occurred. Britain's Aliens Act was enacted in 1905 and was aimed primarily at controlling Jewish migration. Despite discrimination, the Jewish population achieved marked social success in Britain. Many future generations integrated and became successful professionals.

Russian Jews, persecuted by the pogroms, departed to other parts of Europe including Germany and France. Outside Europe, the United States was also a destination. In addition, some Russian Jews left for Palestine. Anti-Semitic legislation was enacted in a number of places. For instance, after the collapse of Béla Kun's communist government, Hungary issued anti-Semitic laws in 1920, 1938, and again in 1939, which restricted the civil rights of Hungarian Jews (Holmes, 1995). After World War I, as Jewish persecution increased in certain countries, it became more difficult for Jewish migrants to find refuge in other countries. In Britain, the 1919 Aliens Act placed restrictions on who could and could not enter the country. In 1924, the United States issued tough legislation to decrease the rate of immigration. Of course, the ultimate anti-Semitic acts were introduced in the early 1930s by the Nazi regime in Germany. While some Jewish refugees were able to migrate to Britain and the United States, millions perished in the Holocaust. Others migrated to France and the Netherlands, only later to perish under the Nazi invasion during World War II (Holmes, 1995).

During the rise of industrialization, there was a large movement of people from rural areas to cities. In Europe, there were 23 cities with a population of more than 100,000 in 1800. This number rose to 125,000 by 1900 (Moch, 1995). Agricultural workers were attracted to the new emerging factories and industries agglomerated in certain regions. In the case of Germany, many agricultural workers from eastern Prussia, mostly ethnic Poles, moved to urban centers of heavy industry and mining in the Ruhr region. By 1913, an estimated 164,000 of the 410,000 Ruhr miners were of Polish background (see Castles and Miller, 2009). As a result, there was a loss of agricultural labor among the large estates of eastern Prussia, and thus the landowners began recruiting "foreign Poles" and Ukrainians. Fear of a takeover of Poles in that part of Prussia prompted the Prussian government to deport some 40,000 Poles and prevent any further in-migration of Poles to the region. The landowners of east Prussia—the Junkers—protested. As a compromise, the government permitted Poles to temporarily migrate as seasonal workers. These temporary migrants were forced to leave each year for a few months, and they were not allowed to bring their families with them.

The "foreign" Poles were not restricted to work in agriculture. They were also recruited to work in industry. In fact, as Castles and Miller (2009, 89) suggest, foreign labor migration was an important part of industrialization in Germany. Along with Poles, there was an in-migration of Italian, Belgian, and Dutch workers. The German government tried to prevent permanent settlement of foreign workers during the early 1900s, a sign of what was to come later when Germany established its "guest worker program" in the 1950s. By 1944, the Nazi regime employed about 7.8 million workers and prisoners-of-war in their factories and elsewhere. At that time, nearly 30 percent of employed persons in Germany were foreigners (Bade, 1995). Foreign labor was an important aspect of Germany's rise as an industrial and military power.

THE GERMAN ECONOMIC MIRACLE

The largest wave of immigration to Germany took place in the 1960s. This was the period during and shortly after the *Wirtschaftswunder*, or "economic miracle," when following the destruction of World War II, there began a rapid reconstruction and development of the German economy. In this postwar period, Germany had a large labor shortage and subsequently signed a series of bilateral agreements to recruit workers from Italy in 1955, Greece in 1960, Turkey in 1961, Morocco in 1963, Portugal in 1964, Tunisia in 1965, and Yugoslavia in 1968. These immigrants to Germany were referred to as "guest workers." From the 1950s until recruitment was discontinued in

1973, some 14 million foreign workers entered Germany and about 11 million returned home. The rest stayed and were joined by their families.

Another major group of immigrants are repatriates of German descent who lived in the former states of the Soviet Union. With the collapse of the communist system of the Soviet Union in the 1980s and early 1990s, many people of German descent moved back to Germany. Owing to the variety of ethnic German repatriates from former Soviet Union states, as well as a more recent increase in asylum seekers from war-torn former Yugoslavia, Iraq, and Afghanistan, the immigrant population diversified greatly. There are currently more than 15 million people with immigrant backgrounds living in Germany, and of these the largest groups include repatriates from countries such as Russia, Poland, and Romania. There are also currently 2.5 million immigrants from Turkey and about 1.5 million from the former Yugoslavia and its successor states.

Over the years, Germany has struggled to integrate immigrants into its society. For a long time, the federal government refused to acknowledge that Germany was a country of immigration. Beginning in the 1990s, political parties from all sides began to grapple with immigration as an important domestic policy issue, especially in light of a number of bloody acts of violence against foreigners. One of the first moves was to introduce legislation, beginning in 2000 with the reformation of Germany's old citizenship law, which allowed a person to acquire citizenship irrespective of place of birth only if their parent was German. The new reformed law allows long-term legal foreign residents and their German-born children the opportunity to acquire German citizenship.

The Independent Commission on Migration to Germany reiterated the need for national immigration reform, urging the implementation of reform designed to counteract shortages in the skilled labor force that are predicted to occur because of the ageing and shrinking German population. Germany's birth rate is among the lowest in Europe and, according to the Commission's report, without immigrants the population would drop to less than 60 million by 2050, and the working population would drop to 26 million from 41 million. The report advocated immigration as a way to help solve the potential shortage of workers.

After much controversy, in 2005 Germany passed the Immigration Act, which for the first time established a legal framework not only to regulate immigration but also to promote the integration of immigrants into German society. The new immigration law helps regulate the inflow and need for skilled foreign workers. A criteria-based point system for the controlled approval of immigrants was established. The 2005 Immigration Act also dealt with the admittance of asylum seekers and, for the first time in German history, introduced mandatory measures to promote immigrant integration. A new, centralized migration and integration administration, the Federal Office for Migration and Refugees, was established and language and orientation courses for immigrants instituted.

Germany's immigration policy reform in recent years signals an important step in the nation's identity as culturally diverse. The policy recognizes the importance of immigrants to the nation's economy as well as the need to better integrate immigrants into German society.

In France, the number of migrants increased rapidly from 381,000 in 1851 to 1 million in 1881 and to 1.2 million, which accounted for about 3 percent of the total population, by 1911 (Weil, 1991). Most of the foreign-born population in France during this time came from the countries of Italy, Belgium, Germany, Switzerland, Spain, and Portugal, and later Poland. By 1931, about one-half of the foreign-born population in France was Polish and Italian (Noiriel, 1995). Most who migrated to France during this period did so for economic reasons, although some of the Italians sought refuge from the Mussolini dictatorship, and Jews fled anti-Semitic persecution in Poland. France desperately needed workers at the time. The fertility rate in France had dropped just as large-scale industrial production emerged. Between 1880 and 1930, there was a major labor shortage in France; French peasants resisted moving to large cities and held tightly to the small family farm. At the beginning of the twentieth century, industrialists and owners of large agricultural enterprises began recruiting miners from Italy and agricultural workers from Poland (Noiriel, 1995).

The two World Wars created much disruption and led to the large-scale displacement of many populations across Europe. Global institutions emerged during this time to address the refugee crises (see Chapter 5). Many of the Europeans who fled Nazi persecution in Germany and occupied Europe during and after World War II fled to the United States and Canada. With the War's end, Europe needed to rebuild. Between 1945 and 1973, the more industrialized countries in Europe used temporary labor recruitment to help expand their economies. The more advanced European economies recruited workers from less developed countries, sometimes through various "guest worker" programs.

In the case of Britain, the government recruited 90,000 male workers from refugee camps and from Italy immediately after World War II. Belgium also recruited Italian men to work in the coal mines and the iron and steel industry. France employed about 150,000 seasonal agricultural workers per year. Most of these workers came from Spain. In the case of Germany, the Federal Labor Office, the Bundesanstalt für Arbeit (BfA), recruited foreign workers from various Mediterranean countries. Germany entered into bilateral agreements with the sending countries of Italy, Spain, Greece, Turkey, Morocco, Portugal, Tunisia, and Yugoslavia. The number of migrant workers in West Germany was 95,000 in 1950, and it increased to 1.3 million by 1966. As we discuss in more detail in Chapter 5, the policy in Germany required that these workers be temporary and be granted labor permits for restricted periods of time. Of course, many of these workers remained and encouraged their families to join them.

Europe eventually became much more integrated economically, a process that began with the Treaty of Paris in 1951 and the Treaty of Rome in 1957. The Treaty of Paris established the European Coal and Steel Community, whose founding members included Belgium, France, Italy, Luxembourg, the Netherlands, and West Germany. Workers in the steel and coal industries could move freely among these nations. The Treaty of Rome established the European Economic Community whose aim was to create a common market among the same six founding members. The common market meant a free movement of goods and workers among member nations. As we explore in Chapter 5, these early treaties and institutions were the foundations for development of the European Union and subsequently a "European Labor Market" among member states in 1993.

In summary, one of the primary reasons for internal migration with Europe during the mid-nineteenth to mid-twentieth centuries was economic. During this time, movements within Europe were partly driven by growing industrialization, the changing nature of agricultural production, and, later, the need to rebuild after World War II. Laborers across Europe shifted to large cities and regions to work in industry. Agricultural enterprises and industry sought out foreign unskilled workers. People also moved for political reasons, and politics certainly influenced the nature of the migration process for particular groups, in particular countries, and especially during times of war. Political tensions over territories and among groups were always fresh in the minds of European legislators seeking to control migration at the time.

MIGRATION AND THE SOUTHEAST ASIAN PERIPHERY

Global economic integration not only impacted migratory processes among core industrial countries of the Americas and Europe, but periphery regions of the world were also affected. In this section, we focus on the movement of Chinese and Indian migrants to Southeast Asia. Between 48 and 52 million Chinese and Indian migrants went to Southeast Asia and parts of the Indian Ocean and South Pacific between 1846 and 1940. Of these migrants, some 29 million Indians and 19 million Chinese settled in other nations. In particular, Indians settled in the British colonies of the day, and many Indian migrants came from the labor-abundant region of Madras. Four million Indians traveled to Malaysia, another 8 million to Ceylon, 15 million to Burma, and about 1 million to Africa, other parts of Southeast Asia, and islands throughout the Indian and Pacific Oceans (Huff and Caggiano, 2007; McKeown, 2004).

Chinese migrants primarily came from Guangdong and Fijian in southeastern China. An important destination for Chinese migrants was Malaysia; about 11 million migrated to the Straits Settlement on the Malay Archipelago. Between 1881 and 1939, the total migration to Malaya averaged 826 immigrants to every 1,000 Malayan residents, which was the highest immigration rate in the world at the time. An additional 4 million Chinese migrated to Thailand, and another 2 million to 3 million to French Indochina between 1846 and 1940 (Huff and Caggiano, 2007; McKeown, 2004).

Southeast Asia provided the much-needed exports to cater to the growing demands of industrial production and commodity consumption in core industrial regions of the world. Resource-abundant but labor-scarce, the region of Southeast Asia (particularly the contemporary countries of Burma, Malaysia, and Thailand) thus became the primary destinations for Chinese and Indian migrants. The colonial authorities of these three countries greatly encouraged in-migration. According to Huff and Caggiano (2007), indentured servitude was not an important factor in the migration process from China and India to the three Southeast Asian countries of Thailand, Burma, and Malaysia. Instead, many migrants paid their own travel costs. Indentured servitude for the Chinese and Indians was much more commonplace for migration to the Caribbean and Latin America.

Similar to transatlantic migration, Chinese and Indian migration to Southeast Asia was heavily influenced by advancements in transportation. By the 1880s, steamships had replaced sail ships as the primary mode of transportation for commodities and people. As a result, shipping costs were greatly reduced. In 1869, the Suez Canal opened, further liberating Asia from what Hatton and Williamson (2005, 129) refer to as the "tyranny of distance." Railroads were developed too, making internal transportation within Asia much easier. Similarly, Asia's integration into the global marketplace was of equal importance. This symbolized the rise of free trade policy in countries such as Japan, Thailand, China, India, and Indonesia.

The costs of moving goods within the periphery and between the periphery and core regions dropped precipitously from 1880 until the beginning of World War I. Consequently, Asian export-oriented regions became more integrated into the global economy. Let us consider the case of the Asian rice market. Studies by Latham and Neal (1983) and Brandt (1985) demonstrate that differences between the prices of goods globally converged. The difference between the price of rice in London and Rangoon fell from 93 to 26 percent in the four decades prior to 1913, which led to an increase in exports. In Burma, Malaysia, and Thailand, exports increased from US$104 million to US$639.6 million (1913 dollars) between 1880 and 1936. Burma and Thailand exported mainly rice, and Malaysia exported tin and rubber. The relative price-boom for exports increased the demand for unskilled labor to the export-rich but labor-scarce Southeast Asian market (Huff and Caggiano, 2007).

THE CHINESE MIGRATION EXPERIENCE

For a long time, China has strictly curtailed both the internal movement of its people as well as any migration into the country, although this is changing. In the past, one of the only significant sources of immigration to China was the "Overseas Chinese," people of Chinese birth or descent who lived outside the People's Republic of China. In the early years after 1949, the government issued various enticements for ethnic Chinese living outside the country to return to their homeland. Although assessing the exact numbers who returned is difficult, it is believed that several million ethnic Chinese have returned to China since 1949. The greatest influx came between 1978 and 1979 when large numbers of ethnic Chinese fled Vietnam as relations between the two countries became strained.

Another wave of immigration to mainland China occurred between the 1950s and 1980s as people from different developing countries entered. During this period, China was seen by countries in Africa and other parts of developing Asia as the leading member of the so-called Third World. To facilitate bilateral relations with other developing counties, China allowed some foreign nationals to enter on academic exchanges, for education and training, as diplomats, and commercial and trade representations.

It was also around this time that the movement of peoples from North Korea into China began, with larger numbers crossing the border during the height of North Korea's famine in the 1990s. Some estimates suggest that there are currently as many as 300,000 North Korean refugees in China. In recent years, China has experienced a wave of immigration from countries such as Vietnam, Burma, Cambodia, and other parts of Southeast Asia, many people crossing the border illegally to work in sugarcane fields, garment factories, and on construction sites. Similar to the in-migration of undocumented immigrants to the United States, the driving force for China is cheap labor.

As China's economy has grown and the country has opened up to the outside world, so foreign immigration has increased, even from parts of the developed world. According to official statistics, in 2005, more than 40 million foreigners entered and exited China and some 380,000 were granted permission to reside more permanently

(Zheng, 2007). Many immigrants entering China move to the major cities of Beijing, Shanghai, Shenzhen, and Guangzhou. As China's economy has boomed, there have been an increasing number of foreign investors from developed countries such as Japan, Singapore, the United States, and parts of Europe who have taken up residence in China. While China is overall a country with relatively low levels of immigration, this is changing as the standard of living in China improves and its economy grows.

In summary, migration during this period of industrialization demonstrated three major patterns. First, there was a large migration of Europeans to the Americas. Second, there was a significant shifting of people internally within Europe. Third, there was a mass migration of people from India and China to countries in Southeast Asia. These three population movements were heavily influenced by the integration of an increasingly global economy centered on the labor needs of rising industrial production in Europe and North America, the political conflict and the aftermath of World War II, and the needs of export-rich but labor-scarce periphery nations in the form of Chinese and Indian migration in Southeast Asia.

MIGRATION AND POSTINDUSTRIAL GLOBALIZATION

In this final part of the chapter, we examine contemporary patterns of migration across the globe, focusing on the 1960s and 1970s and beyond. We emphasize three major trends in contemporary global migration. First, we examine the continuation of migration to the "classical migration" countries of the United States and Australia. Since about the 1960s, a major shift has occurred in the demographic characteristics of migrants entering both these countries. The first migration into both places was by Europeans. Now, many in-migrants to the United States, for instance, originate from Latin America, Asia, the Middle East, and, to a lesser extent, Africa. We focus on the years after 1965, the year when the United States' influential Immigration and Nationality Act abolished the earlier quota system that restricted migration based on national origin. This policy change initiated the demographic shift in U.S. migration.

A similar shift in Australia's migration policy led to in-migration from Asia, the Middle East, and South America.

Second, we consider the rise of migration from different parts of the world to European countries. Europe experienced internal migration during a period of industrial expansion, but today many migrants to Europe come from Asia, Africa, and the Middle East. Yet, this is not to say that internal migration in Europe does not exist. As part of our discussion of migration within Europe, we also consider migratory patterns of people from countries of the former Soviet Union. When the Berlin Wall fell dramatically in November 1989 and the Soviet Union collapsed, emigration from countries in eastern Europe increased remarkably, and there also was increased migration between former eastern European bloc countries.

Third, we discuss migration in the Persian Gulf region. The development of oil production and exports in countries such as Saudi Arabia, Kuwait, Bahrain, and the United Arab Emirates has led to an increasing demand for foreign workers. These oil-rich economic centers attract large companies and the wealthy from many parts of the world but they also require less-wealthy, unskilled migrants who provide domestic and other services to the elites.

MIGRATION TO THE UNITED STATES AND AUSTRALIA SINCE 1965

After a decline in migration from the 1920s until approximately the 1970s, the United States once again became a major destination point for migrants, particularly for people migrating from Latin America and Asia. In 1960, there were about 9.7 million immigrants in the United States. This increased to about 14 million in 1980, then 19.7 million in 1990, 31.1 million in 2000, and 39.9 million in 2010. In 1960, nearly three-quarters of the foreign-born population in the United States was from Europe. By 2008, 80 percent of immigrants came from Asia and Latin America. According to figures from 2009, there were about 10.6 million Asian immigrants in the U.S. As Table 2.1 demonstrates, over 11 million immigrants in the United States are from Mexico, making up close to 30 percent of all the foreign-born population and about 4 percent of the total population of the country in 2011. The countries of India, the Philippines, and China also feature prominently as migrant-sending countries to the U.S.

Table 2.1 Top ten source countries for foreign-born population in the United States, 2011

Country	*Population residing in the U.S.*	*As a percentage of foreign-born population*
Mexico	11,672,619	29%
India	1,856,777	5%
Philippines	1,813,597	4%
China[1]	1,650,411	4%
El Salvador	1,264,743	3%
Vietnam	1,259,317	3%
Cuba	1,094,811	3%
Korea	1,082,613	3%
Dominican Republic	897,263	2%
Guatemala	850,882	2%

Source: U.S. Census Bureau (2011).
Note: [1]Excluding Hong Kong and Taiwan.

Changes in U.S. immigration policy in 1965 meant the abolition of the national origins quota system (see Chapter 5). It was replaced with a structure that focused on criteria such as family reunion and immigrant worker skills. This policy shift opened up the United States to Asian migration. Prior to the 1960s, Asian migrants typically came from Japan or China. However, after the end of the Korean and Vietnam Wars, refugees from Korea, Cambodia, Laos, and Vietnam entered the U.S. in large numbers. During the spring of 1975, shortly after the Fall of Saigon, 125,000 south Vietnamese arrived as refugees. In the late 1970s, the United States accepted refugees known as "boat people"—ethnic Chinese and Vietnamese fleeing Vietnam. By the 1990s, over 1 million refugees from Laos, Cambodia, and Vietnam had settled in the United States. Refugees, as Stephen Castles and Mark Miller (2009, 129) explain, were often "the first link in the migratory chain." Many Asians came to the United States through the family reunion provisions of the 1965 immigration legislation. Later, highly skilled migrants from India, Korea, and Japan arrived in the United States to participate in the labor force in sectors such as technology, healthcare, and finance.

By the end of the twentieth century, the immigrant population of the United States had grown significantly. Although the demographic composition was quite different, the migration trends mirrored the patterns witnessed in the previous century. Many immigrants arrived in the last two decades of the twentieth century, migrating from different places. While the beginning of the twentieth century saw the settlement of European immigrants, migrants that settled at the end of the century generally came from Latin America and various Asian countries. Once again, migrants arrived in large numbers, and they represented a very diverse population in terms of race, ethnicity, and socio-economic status. For example, during the 1990s, 11 million immigrants settled in the United States, which is the greatest number than in any other decade in the nation's history. Thus, at the dawn of the twenty-first century, some 38 million immigrants called the United States home (Camarota, 2007).

One important trend that emerged in the United States was the shift of settlement patterns from the big city to the suburbs. For most of the history of migration to the United States, new immigrants settled in ethnic neighborhoods in the nation's big cities such as Chicago and New York. Neighborhoods such as "Little Italy," "Polishtown," and "Chinatown" were commonplace across many cities. However, this pattern began to change after World War II. The suburbanization of the white middle class also began to attract immigrants. New immigrants bypassed large cities and migrated directly to the suburbs, just like other Americans (Singer, Hardwick, and Brettell, 2008). In fact, approximately 40 percent of recent immigrants entering the country settled directly in the suburbs (see Hanlon, Short, and Vicino, 2009). Today, more immigrants live in suburbs than in big cities (Hanlon *et al.*, 2009).

Similarly, the development of new immigrant gateway cities in the United States stands out as a distinct characteristic. As we explore in Chapter 4, scholars of migration studies use the term "gateway city" to identify and classify cities and regions where immigrants settle. There are several types of gateways, including former gateways, continuous gateways, and emerging gateways. Former gateways once attracted large numbers of immigrants but no longer serve as destination places. Rustbelt cities such as Pittsburgh, Cleveland, and Buffalo are former immigrant gateway cities. Continuous gateway cities once attracted many immigrants and continue to attract immigrants today. New York and Chicago are typical examples of continuous gateway cities. Emerging gateway cities, such as Atlanta, did not attract immigrants in the past but as population growth booms and new economic opportunities develop, these cities are increasingly attractive to new immigrants.

The Australian experience also illustrates the important role that public policy plays in shaping the patterns of migration. In 1901, the Australian government enacted the Immigration Restriction Act, which was "to place certain restrictions on immigration and to provide for the removal from the Commonwealth of prohibited immigrants" (Australian Government, 2013). The Act placed many strict restrictions on who was eligible to immigrate to Australia, such that it served as a de facto "white only" policy. For example, eligibility requirements excluded potential migrants that might be insane, might rely on public welfare, might have a communicable disease, might be dangerous or criminal, or might not read. Accordingly, the Act prohibited virtually all non-white migrants.

During the early twentieth century, the government strongly supported the policy. For example, in 1919, Prime Minister William Morris Hughes argued that the white policy was, "the greatest thing we have achieved." Later, during the early 1940s, John Curtin, then prime minister, staunchly supported the white policy. He stated on record that, "this country shall remain forever the home of the descendants of those people who came here in peace in order to establish in the South Seas an outpost of the British race" (Australian Government, 2013). Similarly, public support also remained strong.

Australia's white policy reflected public attitudes and colonial supremacy of the early and mid-twentieth century. These public policies greatly restricted migration to Australia for many decades, and Australia did not become a fully open society until the later parts of the century. Today, the nation's "Migration Program" permits people to migrate to Australia from any country and does not discriminate on the basis of ethnicity, culture, religion, or language. Remarkably, at the beginning of the twenty-first century, nearly half of Australia's residents reported to the government that they were born abroad or had a parent born abroad. Australia demonstrates that the will to enact public policies can strongly influence migration patterns, whether they restrict or encourage migration.

MIGRATION TO EUROPE

Patterns of migration to Europe in part reflect the history of colonization. Throughout the twentieth century, many migrants of former European colonies settled in the home countries. Two notable trends stand out. First, the United Kingdom is home to many immigrants from the former British colony of India. Many Indians settled in London in the late 1940s and early 1950s—some 60,000 arrived during this period. During the 1960s and 1970s, many more Indians settled in search of employment opportunities. Today, the United Kingdom has some 1.4 million people of Indian ancestry, and they form the largest ethnic minority group in the nation (Spencer, 1997). Second, in France, Algerians are the largest ethnic minority. By 2000, there were 1.7 million people of first- or second-generation Algerian descent, which accounted for approximately 14 percent of the immigrant population (Institut National de la Statistique et des Études Économiques (INSEE), 2012). Following the conclusion of the Algerian War in 1954, a quarter of a million Algerians migrated to France. Later, upon independence from France in 1962, many more Algerians migrated to France, particularly to metropolitan Paris (Singer, 2013).

In Germany, Turks make up the largest ethnic minority in the nation. Approximately 1.6 million Turkish immigrants live in Germany, and another 4 million people have at least one parent that was a Turkish immigrant. Beginning in the 1960s, Turks began to migrate to Germany to participate in the nations' guest worker program. A lack of employment and economic opportunities in Turkey prompted many Turks to migrate elsewhere. Later, many more Turks arrived to reunite families (Mandel, 2008).

Internal migration within continental Europe is also a feature of European migration. Since the fall of the Berlin Wall in Germany and the collapse of the Soviet Union, the predominant migration pattern has been the movement of people from eastern Europe to southern and western Europe. For example, since the 1980s, many eastern Europeans settled in other parts of Europe, including: Poles migrated to the United Kingdom; Czechs migrated to Ireland; Romanians migrated to Italy and Ireland; and Serbs migrated to Spain and Italy.

These migration patterns have had three important consequences for Europe. First, the European population is significantly more diverse, and this diversity is increasingly varied. Racial and ethnic diversity is more pronounced today in the United Kingdom, France, and Germany. While these nations were historically home to monolithic populations, they now reflect a diversity of people from the latest migration patterns from their colonies. Second, the history of European colonization is closely tied to the industrialization of these nations. The rise of manufacturing economies, particularly in the production of goods from raw materials, necessitated a large pool of low-skilled labor. In many instances, European nations welcomed the migration of people from their former colonies and other places to satisfy the demand for this labor force. Third, the rise of the global economy prompted a marked shift in migratory patterns to Europe. A global economy that is largely based on services such as health, banking, and technology created additional demands for Europe to compete for a new high-skilled labor force. Thus, a new generation of educated and skilled migrants settled in Europe.

Like in many other nations, debates about citizenship continue on an ongoing basis throughout Europe. Today, migration flows happen seamlessly as a result of agreements among nations in the European Union. Nonetheless, European nations, their societies and economies, have been transformed by the arrival of new migrants.

MIGRATION TO THE PERSIAN GULF REGION

Migration has shaped the urbanization and economic growth of the Persian Gulf region. Let us consider the United Arab Emirates. Dubai, a large metropolitan area, doubled its population between 1995 and 2005, to 1.2 million residents. This growth is attributed to the economy of the oil industry, and later in the mid-1990s, the emirate adopted a new business model that emphasized diversifying the economy and reducing dependence on oil as the primary source of economic growth. Dubai established itself as a major tourist destination and financial center, a place of "free zones" where multinational corporations conduct business tax free, and where there now exists an impressive skyline of new hotels, condominiums, malls, and theme parks. Dubai exploded with new real estate, and while the property economy of Dubai suffered tremendously in the recent global financial crisis, the building boom created some of the tallest skyscrapers and largest real estate projects in the world (Krane, 2009).

In 2005, the foreign-born population of Dubai was over 1 million people, 83 percent of the total population. About half of these immigrants came from India. Of the total population, about 85 percent are employed, and of the employed population, 97 percent are foreign labor. Nowhere is the story of a close relationship between the economy and immigration more obvious than in the story of Dubai. Immigrant workers from poor villages in India, Pakistan, and Sri Lanka fueled the building boom in Dubai. The push factors of dismal prospects in the impoverished home countries have encouraged workers to seek new opportunities. At the same time, the demand for cheap labor in Dubai has been a considerable pull factor for migration into the city.

Yet, for the immigrant workers, it can be a rather bleak situation. Paid meager wages and working under harsh conditions, immigrant workers involved in construction in Dubai have become essentially indentured servants. Many are crowded into employer-provided housing in the form of trailers sitting behind barbed wire on the fringes of Dubai's desert. Some immigrants borrowed heavily to pay recruitment agents for their jobs and are stuck paying back debts that could take several years to pay off because of low wages. In Dubai, employers often confiscate the passports of their workers and withhold pay for months on end. As the case of Dubai suggests, the sweat of immigrant workers can improve a country's economy. However, in many cases, these workers are badly exploited as cheap labor (Ali, 2010).

CHAPTER SUMMARY

The process of global migration is as old as humankind. In this chapter, we identified broad geographic patterns of migration over the past few centuries. Some key themes shaped global migration: Europe's early colonial expansion; industrialization expansion; and globalization.

Between the seventeenth and nineteenth centuries, Spain, the Netherlands, Portugal, Britain, and France established many colonies on various continents, including Asia, Africa, the Americas, and Oceania. This colonization was accompanied by voluntary and involuntary forms of migration. Free people migrated. Slaves were forced to migrate. Indentured servants were forced to migrate under poor conditions. As a result, there was a large outflow of population from Europe to parts of the colonized world. Slaves from Africa were forced to migrate to the Americas. While war and poor health killed many migrants, the migrants that settled had many impacts on their new homes in terms of economy, politics, and culture.

Second, we examined how industrialization expansion led to three notable movements of people from about the mid-nineteenth century to the mid-twentieth century. During this period, industrialization was the dominant characteristic that fueled economic growth. In turn, this created a push and pull force that attracted immigrants from many parts of the world to settle in industrializing nations. Many Europeans migrated to the Americas, and there was also a significant resettling of people within Europe. Also, there was a mass migration of people from India and China to countries in Southeast Asia. These population movements did not occur accidently; rather, they were influenced by industrialization, globalization, and political conflict.

Last, we explored various contemporary patterns of migration since the 1960s. Three trends are notable. First, there has been a continuous pattern of migration to the "classical migration" countries of the United States and Australia. A major shift occurred in the demographic characteristics of migrants during the post-World War II era. While the first pattern was primarily migration from Europe, the second pattern included migrants originating from Latin America, Asia, the Middle East, and Africa. Public policies influenced changes in the migratory patterns. In the United States, the Immigration and Nationality Act of 1965 abolished a quota system that restricted migration that was based on one's country of origin. Similarly, the abolition of Australia's "white only" migration policy led to the in-migration from other continents. In Europe, the industrial expansion of the twentieth century led to migrants arriving from former colonies such as India and Algeria, and later, from Asia, Africa, and the Middle East. The collapse of the Iron Curtain also influenced the migration of people from eastern Europe to western and southern Europe. In the Persian Gulf region, the growth of big cities in oil-rich countries attracted multinational companies and wealthy migrants from many parts of the world, as well as unskilled migrants.

Patterns of global migration have been shaped by a convergence of many societal changes in the world. The fall of colonial powers, the rise of industrialization, and the emergence of globalization fueled migration in the twentieth century. Few places in the world have not been affected by migration. Indeed, migration touches the lives of many of the world's population—rich and poor, healthy and sick, free and restricted. People are motivated to migrate for a host of reasons, among them to seek better economic opportunities, flee war, and reunite with families.

GUIDE TO FURTHER READING

There are many great works on the history of global migration. The following offer a comprehensive history of migration spanning the regions of the world:

Roger Daniels (2002). *Coming to America: A History of Immigration and Ethnicity in American Life, 2nd Edition*. New York: Harper Perennial.

Diana Lary (2012). *Chinese Migrations: The Movement of People, Goods, and Ideas over Four Millennia*. Lanham, MD: Rowman & Littlefield Publishers.

Abdoulaye Kane and Todd H. Leedy (Eds) (2013) *African Migrations: Patterns and Perspectives*. Bloomington, IN: Indiana University Press.

Klaus Bade (2003). *Migration in European History*. Malden, MA: Blackwell Publishing.

Samuel L. Baily and Eduardo José Miguez (Eds) (2003). *Mass Migration to Modern Latin America*. Lanham, MD: Rowman & Littlefield Publishers.

MIGRANTS AND SOCIETY

Migration has tremendous impacts on the societies of both sending and receiving countries involved in the migration process. In this chapter, our primary focus is on the relationship between migration and the receiving society. In particular, we consider issues related to ethnicity, gender, and sexual orientation. Migration, especially when it occurs at a significant scale, can change the demographic composition of receiving countries, especially within specific cities, regions, or neighborhoods. The ethnic, cultural, and socio-economic characteristics of migrants are typically very different from those of the native population of the receiving place. These differences can lead to tensions between groups, but they can also lead to the development of new, more interesting, and diverse societies.

The first section of this chapter explores the relationship between migration and ethnicity. Here, we aim to provide the reader with an understanding of ethnicity and how it is related to a host nation's reactions to migration. Then, we examine five major trends in the social aspects of migration. First, we examine the ways that migrants assimilate into the receiving society. As part of this discussion, we consider how the theory of assimilation has changed since it was first formulated as well as its relevance today. Second, we explore the notion of multiculturalism and how this influences ideas about migrant incorporation into society. As part of this discussion, we examine the concept of diaspora and the more recent focus on the notion of transnational migration. Transnational migration refers to the process whereby migrants create multiple social, economic, and political relations that link them to both their destination society and their society of origin. Third, we examine the relationship between migration and security, especially in light of recent terrorists attacks in Europe and the United States. Fourth, we consider the nexus between migration and gender and examine the recent rise in female migration. Finally, we reflect on issues of sexuality and migration, specifically examining historic and more recent reactions to the migration of lesbian, gay, bisexual, and transgendered (LGBT) people.

MIGRATION AND ETHNICITY

Ethnicity is defined as a sense of group belonging based on common ancestry, common language or culture, shared experiences and values, and mutual religion (Plummer, 2010). When migrants move to a new society they often bring with them a sense of belonging to a different culture, with a different set of values, a distinct ancestry, and, sometimes, a different set of religious beliefs to the host society. Migrants belong to certain ethnic groups. Because of migration, societies across the globe retain cultural patterns rooted in many different parts of the world and connected to very different ethnic groups.

For example, the United Kingdom is composed of people who can trace their roots to Ireland, Pakistan, Bangladesh, China, the Caribbean, and many other places. These people connect, to varying degrees and with varying levels of intensity, to others who share the same ancestral home. In some cases, people connect on the basis of an ethnicity that goes beyond national identity. For instance, Kurds living in Turkey connect to other Kurds, and in some ways they see themselves as a separate ethnic group distinct from other people in Turkey. Nations can be composed of different ethnicities. There are instances when ethnic groups self-identify as such; but in other cases, ethnicity can be imposed upon people by powerful outsiders or a dominant paradigm (Jenkins, 1997). Sometimes a group can, with time, identify with a category that is imposed upon them. Ethnicity can become internalized, and people can assume an imposed identity. This can happen to migrants and their children, sometimes in different ways. For instance, children of migrants from Jamaica living in the United States may begin to self-identify as "black" while their parents continue to identify as Jamaican. Or, and most likely, an immigrant family living in the United States may want to be identified as Jamaican, but the tag of "black American" is imposed upon them (Waters, 1999).

Many scholars have attempted to theorize ethnicity. Some have suggested that membership of a particular ethnic group is something one is born into and to which most people have a *primordial attachment*. A proponent of the theory of primordial attachment is anthropologist Clifford Geertz. Geertz (1963, 107) states:

> By a primordial attachment is meant one that stems from the "givens"—or, more precisely, as culture is inevitably involved in such matters, the assumed "givens"—of social existence: immediate contiguity and kin connection mainly, but beyond them the givenness that stems from being born into a particular religious community, speaking a particular language, or even a dialect of a language, and following particular social practices. These congruities of blood, speech, custom, and so on, are seen to have an ineffable, and at times overpowering, coerciveness in and of themselves. ... The general strength of such primordial bonds, and the types of them that are important, differ from person to person, from society to society, and from time to time. But for virtually every person, in every society, at almost all times, some attachments seem to flow more from a sense of natural—some would say spiritual—affinity than from social interaction.

In this sense, members of a particular ethnic group are in some sense "naturally" bound together.

Other scholars disagree with this assertion, and in general, this viewpoint has been increasingly marginalized (Sezneva, 2013). Fredrik Barth (1969) advances a very different viewpoint in his study, *Ethnic Groups and Boundaries*. Here, he suggests that ethnicity is a by-product of historical, political, and social process, not a given but rather defined by social situations and settings. Within this context, social anthropologists emphasize the concept of *situational ethnicity*, which suggests that actors invoke their ethnicity as a defining criterion depending on whether it is advantageous or not to do so in a given social situation. These scholars point out that it may not be advantageous at times to be associated with a particular ethnic group if that group holds a subordinate status in society (Okamura, 1981).

Other scholars suggest that ethnic mobilization or ethnic identification occurs in ways that aim to maximize the power of that group within the competitive market structure. In this sense, attachments to an ethnicity are ascribed and embedded in market relations and socio-economic structures (Castles and Miller, 2009). In a recent book, *Ethnicity Incorporated*, John Comaroff and Jean Comaroff (2009, 15) suggest that attachment to ethnic identity has, for some groups, given them "privileged access to markets, money and material enrichment." This book focuses, in part, on ways corporations can co-opt ethnic markers and practices to promote certain forms of consumption. Markers have always been important in discussions of ethnicity. Recognizable traits of migrants such as language, religion, skin color, name, and other qualities can signal either acceptance or exclusion by the host society.

Much of the debate among scholars has focused on trying to understand why ethnicity is such a prominent and significant category for people. Belonging to a particular ethnic group can be a great source of pride for people. This sense of pride can motivate people to join political and social movements, some with extremist viewpoints and intentions (Couton, 2013). The most extreme example is probably the far-right nationalist movement, or "white pride" movement, that advocates the elimination of other groups that they deem to be inferior. However, ethnic pride can be a source of influence for much less extreme movements, and in a democratic system, politicians can appeal to ethnic pride for votes and in the promotion of specific policies that might benefit particular groups.

It is not unusual for immigrants to reawaken and express their attachment to a particular ethnic group or nation. In the case of the United States for instance, ethnic pride manifests itself as pride among Irish Americans, especially on St. Patrick's Day, or Italian pride, or other forms of ethnic pride depending on the group (Couton, 2013). Writing about this in the 1970s, Herbert Gans (1979) suggested that the renewed expression of ethnic pride among third-generation migrants could be understood as symbolic or affective ethnicity; not necessarily an indication of a lack of migrant incorporation. More recent immigrant expressions of ethnic pride can, at times, be misinterpreted by native populations as a rejection of the destination society, and can lead to ethnic tension. In some instances, depending on the political times, ethnic groups might need to hide their ethnicity. For instance, German immigrants in the United States had to deny their ethnicity during the two World Wars. But in most democratic nations, there is a recognition of the possibility that separate cultures can coexist and live side by side.

According to John Rex (1994), there are three possible reactions of a receiving nation to the in-migration of different ethnic groups. One reaction is xenophobic and racist, in which the nation responds by demanding the expulsion of migrants and offers immigrant fewer rights and privileges than the native population. The second reaction is assimilationist. In this case, ethnic migrants receive equal rights, the same as citizens of the nation, but they are discouraged from maintaining their own separate culture and ethnic identity. For example, France has generally taken this approach toward migrant incorporation, and the United States remains the primary example used by scholars to demonstrate the assimilation process. The third reaction is one of multiculturalism, whereby the ethnic migrant identity is allowed to flourish. European nations have strongly debated a multicultural approach to migrant incorporation. In the following sections, we explore the notions of assimilation and multiculturalism.

ASSIMILATION

Robert Park and other renowned sociologists of the "Chicago School" studied Chicago's neighborhoods during the 1920s and 1930s in an attempt to document how cities evolve and develop, (Park, Burgess, and McKenzie, 1925). As part of this work, these sociologists studied how immigrants and their children become part of a host society. Robert Park and Ernest Burgess, in 1921, proposed a theory of assimilation. Park and Burgess (1967, 735) provided the following definition:

> [Assimilation is] a process of interpenetration and fusion in which persons and groups acquire the memory, sense and attitudes of other

> persons and groups and, by sharing their experience and history, are incorporated with them in a common cultural life.

Over time, ethnic migrant minorities become part of mainstream, in this case, American life. Later, Robert Park (1930, 281) offered another definition in the *Encyclopedia of the Social Science*, stating that social assimilation is "the name given to the process or processes of which people of diverse racial origins and different cultural heritages, occupying a common territory, achieve a cultural solidarity sufficient at least to maintain a common national existence." The theory suggests that over time immigrants assimilate and become more like the native population of their new destination. While Park and others identified concerns such as poor housing and a lack of adequate working conditions among immigrant communities, they also found that over time immigrants learned the language and norms of the native population in ways that allowed them to become more integrated into the host society (Powers, 2013).

Some 30 years later, in his book *Assimilation in America*, Milton Gordon (1981) amended the earlier model of assimilation. In his book, he identifies at least three stages of assimilation. The first stage is cultural assimilation where immigrants adopt the language and customs of the host society. This defines acculturation. The second stage is structural assimilation where immigrants become integrated into schools, employment, and various friendships. The final stage is intermarriage. Here, immigrants truly assimilate by marrying members of the host society and having children with them (Powers, 2013).

For Milton Gordon, acculturation or cultural assimilation could occur without structural assimilation, or structural assimilation could take place at a later stage (Alba and Nee, 1997). Some scholars have argued for "straight-line assimilation." Popularized by Herbert Gans (1979), straight-line assimilation suggests that assimilation unfolds over generations. The first immigrants arrive, they have children, and, with each subsequent generation, assimilation unwinds and is completed in a somewhat linear fashion. Later, Gans (1992) admitted that assimilation of immigrant groups does not always follow a straight line. Indeed, there can be bumps along the way. The pace of assimilation may vary.

Milton Gordon (1964) was aware that, implicit in the theory of assimilation, is the notion of a dominant culture. He states (1964, 5), "The white Protestant American is rarely conscious that he belongs to a group at all. He inhabits America. The others live in groups. One is wryly reminded of the comment that the fish never discovers water." A pertinent criticism of cultural assimilation suggests that it is ethnocentric, stipulating that minority migrants abandon their ethnic and cultural identity to become part of the majority culture. Cultural pluralists reject this process. Others (Alba and Nee, 1997, 834) have argued, "acculturation need not be defined simply as the substitution of one cultural expression for its equivalent. ... The influence of minority ethnic cultures can occur also by expansion of the range of what is considered normative behavior in the mainstream." In other words, minority cultures can heavily influence the majority culture during the process of acculturation such that cultures become fused to create a "hybrid cultural mix" (Alba and Nee, 1997, 834).

As Milton Gordon (1964) pointed out, culture is only one dimension of assimilation. Since the theory of assimilation first emerged, scholars have been concerned with socio-economic assimilation. One aspect of socio-economic assimilation relates to social mobility (Alba and Nee, 1997). In many cases, the socio-economic conditions and labor market earnings for immigrants from ethnic communities initially differ from those of the host population. But socio-economic assimilation theory suggests that, although most migrants enter at the bottom of the labor market, migrants and their descendants achieve upward mobility such that, over time, they reach at least an equivalent social standing to that of the host generation.

As we explored in Chapter 2, migrants entering the United States, Canada, and Australia during much of the late nineteenth and early twentieth centuries were mostly White Protestants from northern and western Europe. Later, migrants from southern and eastern Europe, often Catholic or Jewish, began entering and settling in large industrial cities, especially in the United States. Figure 3.1 is a contemporary photograph of Little Italy in New York, historically an Italian immigrant neighborhood that emerged as part of this type of migration. Later still, the migrant population to these same developed nations became more diverse, coming

Figure 3.1 Little Italy in New York.
Source: Erlend Bjørtvedt, available online at http://commons.wikimedia.org/wiki/File%3ANY_Little_Italy_ IMG_2059.JPG (accessed June 25, 2013).

from places in Latin America, Asia, the Caribbean, and Africa, especially after the 1960s. The range of ethnicities entering classical immigrant nations expanded and ethnic communities evolved.

Certainly, upward social mobility occurred for many of the European migrants who entered the United States, Canada, and Australia during the early twentieth century. Many scholars agree that these migrants achieved socio-economic assimilation, some suggesting that the opportunity structure was such that European migrants, in particular, to the United States could achieve a level of success. As Mary Powers (2013, 4) suggests:

> The economy of the 1940s–1970s was expanding through mainstream jobs that provided good wages, salaries, and benefits. The post-World War II era was also a period of relatively low immigration, low unemployment, and increasing educational attainment, which provided opportunities for advancement and security for the second generation of white European immigrants.

The more recent debate on assimilation theory questions the reality of socio-economic assimilation for the generation of diverse migrants who have entered classic immigrant nations over the past few decades. This brings us to another variant on socio-economic assimilation, segmented assimilation. In a classic article, Alejandro Portes and Min Zhou (1993) argue that second-generation contemporary non-white migrants could take several paths toward incorporation. They can incorporate into the white mainstream American society, or alternatively they can incorporate into the non-white underclass. Assimilation is segmented for more contemporary immigrants such that they can be subjected to "downward mobility" (Powers, 2013).

Beginning in the 1970s, structural shifts in the economy of developed nations meant the loss of many well-paid industrial jobs. Today's immigrants and their children are confronted with a very different economic situation than earlier European migrants. They can expect fewer job opportunities, and available employment is much more precarious. Employment opportunities for many migrants are often in the form of part-time or temporary contract work that is poorly paid and offers few benefits (Powers, 2013). According to Alejandro Portes and Rubén Rumbaut (1996), labor migrants, particularly Mexicans, are most impacted by economic restructuring. Combined with this reality are issues of race, especially in the context of the United States. Racial and ethnic discrimination negatively affects the immigrant experience and the possibility for upward mobility, especially if the immigrant is black (Waters, 1999).

The traditional immigrant story in the United States is one where new arrivals join friends and family, living in clustered neighborhoods in the city. Gradually, with more contact with the new culture and customs, they lose connections with their home country, create new social ties, adopt the language and ways of the receiving nation, disperse and achieve the dream of success and homeownership in the suburbs. There is an obvious spatial component to this type of assimilation. Spatial assimilation links the importance of residential mobility to the assimilation process. As migrant ethnic groups assimilate both culturally and socio-economically, they leave behind their traditional ethnic community, and they translate their socio-economic success by purchasing a home in a community with better amenities and more advantages (Alba and Nee, 1997). Typically, in the U.S. context, this means they move to the suburbs. One new and interesting feature of contemporary migration to the United States is that many immigrants are bypassing the inner city and in-migrating directly to the suburbs. In light of this, the question remains whether suburbia can continue to be seen as the location of immigrant success and a place of advantage.

SUBURBAN IMMIGRATION IN THE UNITED STATES

Silver Spring, Maryland, is a suburb of Washington, DC. Located to the north of the nation's capital, the suburb initially grew during the 1960s and 1970s as a bedroom community. Today, approximately 73,000 residents live in Silver Spring, and the population has dramatically changed since its roots as a Washingtonian suburb. In 1980, three-quarters of the population was white; yet three decades later, less than 40 percent of the population reflected this historical pattern. In 2010, approximately 45 percent of the population was white; 28 percent was black; 8 percent was Asian; and 18 percent represented other races. A quarter of the population was Hispanic or Latino. What is even more indicative of these changes is that

37 percent of the population was foreign-born. During these 30 years, the national suburban population mirrored this transformation of Silver Spring.

There is no doubt that the United States is a nation of suburbs. Two-thirds of Americans live, work, play, and vote in suburban communities that are located in metropolitan areas but outside of the nation's large central cities. Indeed, the stereotypical suburb as a white, middle-class residential community no longer portrays today's contemporary suburb. However, the suburban transformation occurred very slowly until after World War II (Fishman, 1987). A brief look at the historic patterns of immigration of the United States shows that the country is also a nation of immigrants. Prior to the immigrants settling in suburbs, they settled in many different regions in distinct spatial patterns. In the first wave of immigration, the earliest immigrants arrived during the seventeenth and eighteenth centuries from England. Arriving in the northeastern states, they settled and called their land "New England." The American Colonies were born. Today, the region is composed of six states: Maine, Vermont, New Hampshire, Massachusetts, Rhode Island, and Connecticut. Then, in the second wave of immigration, from approximately the 1820s to the 1880s, millions of Irish and German immigrants settled not only in New England but also in more rural regions of New York, the Mid-Atlantic, and the Midwest regions. Later, the third wave of immigration brought over 20 million immigrants from eastern and southern Europe to the United States. From the 1880s to the 1920s, these immigrants passed through famous ports such as New York's Ellis Island to settle in big cities along the East Coast and Midwest. Recently, the current, fourth, wave of immigration has witnessed new immigrants from Latin America and Asia settling in metropolitan areas across the country—big and small—and in cities and suburbs.

In the last decade of the twentieth century more immigrants entered the United States than in any other decade in the nation's history. It is estimated that the immigrant population, both documented and undocumented, reached a record 37.9 million in 2007 (Camarota, 2007). In the 1990s alone, 11 million immigrants entered the United States. Traditionally, new arrivals to the U.S. in-migrated to ethnic neighborhoods in such large cities as New York, Chicago, and Boston. In a study of the nation's 18 largest regions, Vicino, Hanlon, and Short (2007) found that 25 percent, or 9.8 million immigrants, lived in dense, urban neighborhoods in large central cities. A classification of the population's primary characteristics included Hispanic, white working class, gentrified, and Asian. The urban centers of New York, Los Angeles, San Francisco, and Chicago stood out as the largest cities where the greatest number of urban immigrants settled.

In contrast, this traditional settlement pattern has changed in a dramatic way. By the beginning of the twenty-first century, many new immigrants began to bypass the central city completely and migrated directly to the suburbs. According to a recent report by the U.S. Census Bureau, in 2006 four in every ten immigrants settled directly in the suburbs, sidestepping the traditional urban core (Roberts, 2007). In particular, many new immigrants settled in three types of immigrant gateway suburbs (Singer *et al.*, 2008). First, "emerging gateways" are regions where the suburbs of Atlanta, Dallas, Las Vegas, Orlando, Phoenix, and Washington, DC, have grown at a very fast pace. In 1980, the foreign-born suburban population was just over one and a half million residents, but, by 2005, about 3.9 million immigrants called the suburbs of these regions home. Second, "re-emerging gateways" include the suburbs of Denver, Minneapolis, Portland, Sacramento, San Jose, Seattle, and Tampa, where there was a larger immigrant population base in 1980, but growth later occurred at a slower pace. In 1980, approximately 721,000 immigrants lived in the suburbs of these regions, but, by 2005, over 2.5 million immigrants had settled here. Third, the "pre-emerging gateways," of Austin, Charlotte, Greensboro, Winston-Salem, Raleigh, Durham, and Salt Lake City had barely 90,000 immigrants living in their suburbs in 1980; three decades later, 670,000 immigrants had settled there. In sum, researchers have identified 20 immigrant gateway suburbs. In 1980, there were 1.3 million immigrants in these regions, and, by 2005, 7.1 million immigrants became suburbanites. It was clear at the end of the twentieth century that there would be more immigrants living in suburban America than in the nation's central cities. This geographic shift demonstrates that immigrants, just like native U.S. residents, have many things in common: particularly the American suburban way of life.

This new geographic pattern of immigration has important implications, particularly in the realm of public policy. At the level of local government, policies and ordinances aimed directly at a city's immigrant population have emerged as significant elements of the immigration debate. Some cities have declared themselves "sanctuary cities," following practices that protect the immigrant population. In these cities, which include New York, Washington, DC, San Francisco, Detroit, Miami, Portland, Seattle, Dallas, Houston, Portland, and Chicago, police and other local government agencies refuse to provide information on the legal status of local immigrants to federal authorities. In contrast to the cities, many surrounding suburbs as well as other smaller cities in the U.S. have adopted policies that are designed to exclude undocumented immigrants from their local jurisdictions. As the federal government falters in developing an adequate national immigration policy, localities have adopted a variety of policy positions and approaches to the issue. For example, while immigrants might be "accepted" and "protected" in the city of Washington, DC, immigrants, at least undocumented immigrants, can be "rejected" by neighboring Prince Williams County, a suburban county in the Washington, DC area. As Varsanyi (2008, 892) states, this is "literally creating a patchwork and layered geography of personhood and alienage."

Without a doubt, the United States is a "suburban nation." As this case illustrates, the change in the demographic characteristics of residents is noteworthy. Suburbs such as Silver Spring, Maryland are now gateway suburbs. They have emerged as the new hub of metropolitan regions where immigrants settle upon arrival in the country—completely bypassing the historic central city. Social, economic, and political tensions arise in suburbs in the same way that they did in big cities a century ago. Suburbs are the new contested spaces in which immigration policy in the U.S. will be decided.

Overall, assimilation theory has been widely criticized in recent years, especially in light of new migration patterns that emerged in developed nations after the 1960s. The effects of economic restructuring and racial discrimination are thought to negatively impact the ability of these new migrants to achieve upward mobility. In a large-scale study of unemployment levels among second- and third-generation migrants of European and non-European ancestry in countries such as Australia, Canada, Israel, South Africa, the United States, and a variety of countries in Europe, Anthony

Heath and Sin Yi Cheung (2007) found second-generation Europeans did not experience an "ethnic penalty" when compared with non-Europeans. In other words, employment rates among second-generation Europeans were the same as or better than those of the native population. In all countries, non-Europeans experienced the opposite. According to Heath and Cheung (2007), reasons for the differences include racial discrimination and contemporary labor market flexibility. Still, there are some scholars who suggest that assimilation theory continues to bear weight in attempting to explain migrant integration in the United States (Alba and Nee, 1997); others suggest that assimilation is segmented (Portes and Zhou, 1993); and still others look for new models of immigrant integration. In the following section, we examine one such model: multiculturalism.

MULTICULTURALISM AND TRANSNATIONAL MIGRATION

In 1908, Israel Zangwill's play *The Melting Pot* popularized the term that is still today used to describe the assimilation process in the United States. The "melting pot" metaphor suggests that cultural differences among ethnic groups "melt away," or disappear, as the migrant assimilates and becomes Americanized. As a contrasting metaphor, the term "salad bowl" has been used to signify multiculturalism. Many different cultures maintain their distinctive elements and yet coexist together in one multicultural society. With multiculturalism, cultures do not melt away to merge into a homogenous society; rather, these cultures remain celebrated and distinct.

As Natalia Mehlman Petrzela (2013) suggests, multiculturalism can be descriptive. The term can be used to describe a society. For instance, the United Kingdom is a multicultural society. Alternatively, multiculturalism can be normative, a projection of a "political aim of recognizing group identity—usually ethno-racial or cultural, but increasingly gender, sexual, and religious—in the public sphere, and crafting policy accordingly" (Petrzela, 2013, 1). By the end of the 1960s, multiculturalism became a popular approach to public policy. As such, it signals a commitment by the state to ensure the equal rights of ethnic migrant groups. It also acknowledges and accepts their ethnic differences. As we will explore later, some politicians and policymakers have, at least publicly, backed away from multiculturalism in more recent times, especially after September 11, 2001 and in the midst of growing concern about Islamic fundamentalism (Castles and Miller, 2009).

Multiculturalism has its historical roots in a number of important developments beginning shortly after World War II. First, reactions to the mass genocide of Jews and others during the war led to the evolution of human rights as well as a repudiation of the scientific racism endemic in the Nazi regime. The idea that particular races or ethnicities were inherently inferior—already on the wane—was largely rejected. The inherent equality of all human beings regardless of ethnicity or race was embraced. Second, the fight against racial and ethnic discrimination was embraced by the civil rights movement of the 1960s in the United States. This led to a demand for minority rights and the encouragement of pride among minority cultures. Race was explored as a social construct rather than a biological marker. The legacy of any sort of racial or ethnic hierarchy was rejected in favor of an ideology of equality among races, and the notion that each racial and ethnic group could make a unique contribution to society (Kymlicka, 2012).

Theoretically, contemporary proponents of multiculturalism call for a "politics of recognition" (Taylor, 1994) where, as Natalia Mehlman Petrzela states (2013, 3), "there is no dominant culture to which different groups must assimilate, but rather a set of public institutions which recognize the equal worth of the various groups, and officially supports their practice and development." The cosmopolitan viewpoint stresses the importance of shared values across different groups and recognizes that there should be a set of basic principles to which we all abide, regardless of our culture (Petrzela, 2013).

Keith Banting and Will Kymlicka (2006) developed a Multicultural Policy Index (MPI) to monitor the development of multicultural polices across 21 different nations. As part of this work, they categorized eight types of policies as being common across these different countries. They include policies related to the adoption of multiculturalism in schools; the inclusion of different ethnicities and races in the public media; exemptions from certain dress codes; allowing dual citizenship; funding for bilingual education; funding for ethnic groups to support cultural activities;

affirmative action policies; and constitutional or legislative affirmation of multiculturalism (Kymlicka, 2012). In an analysis of the MPI over time, Banting and Kymlicka found that, despite the backlash against multiculturalism in more recent years, multiculturalist policies have continued to expand over the past few decades.

Some of these policies have sparked controversy. An example is the dispute centered on the wearing of a headscarf (or hijab) by Muslim girls and turbans by Sikh boys in French schools. In 2004, France adopted a law banning the wearing of any religious symbols in French public schools. Proponents of the law saw this as an important step to affirming the principle of secularism, a defining value of the French state. Those against the law perceived it to be discriminating against Muslims. This served to only energize anti-Islamic feelings in France (Koser, 2007).

Some multicultural practices, particularly legislative measures, focus on the "equal rights" aspect of multiculturalism and reject the notion that ethnic difference can be the basis for deprivation of any social goods, such as housing or education. Others are more "color-conscious" (Petrzela, 2013). Examples of the latter include bilingual education or other services targeted to help a particular ethnic group or set of ethnic groups. Affirmative action falls into the latter category. It is often the "color-conscious" policies that are the most controversial.

Critics of multiculturalism include people on the "left" and "right" of the political spectrum. Those that are more "left-leaning" are concerned that multiculturalism, by celebrating ethnic differences, is splintering the working class. Others have deep concerns that multiculturalism ignores larger socio-economic inequalities and conditions of political isolation experienced by many migrant ethnic minorities. As Will Kymlicka (2012, 4) suggests:

> Even if all Britons come to enjoy Jamaican steel drum music or Indian samosas, this would do nothing to address the real problems facing Caribbean and South Asian communities in Britain—problems of unemployment, poor educational outcomes, residential segregation, poor English language skills, and political marginalization.

It is not enough to merely focus on the promotion of cultural diversity. Efforts must be made to solve socio-economic problems among ethnic migrants and their families.

Some generally more conservative groups worry that multiculturalism leads to a loss of national identity. Samuel P. Huntington (2004), for example, suggests that embracing Mexican migration and ethnic identity is a threat to American culture and identity. Those who advocate for more exclusionary policies toward migrants typically cite concern that increasing ethnic diversity threatens national identity. For these groups, acceptance of national identity can only occur with rejection of prior nationality. According to Kymlicka (2012) if an immigrant to Germany or France has excessive attachment to his or her nation of origin, they can be refused naturalization. It is difficult for some to accept that a migrant can hold two or more identities.

Yet, researchers suggest that there is rapid growth in the number of people with affiliations and strong attachments to different countries. The term transmigrant has been used to describe migrants who live "in-between" nations, maintaining strong social, economic, and political ties with their nation of origin while living in their host destination. Alejandro Portes (1997, 821) describes transnational communities as

> [d]ense networks across political borders created by immigrants in their quest for economic advancement and social recognition. Through these networks, an increasing number of people are able to live dual lives. Participants are often bilingual, move easily between cultures, frequently maintain homes in two countries and pursue economic, political and cultural interests that require presence in both.

Tied to this notion of transnational migration is the concept of the diaspora. The term diaspora is a classical term used in reference to large-scale displacement of people in earlier times. An example is the Jewish diaspora following the destruction of the Second Temple in 586 BC (Koser, 2007). More recently, some theorists suggest that there are contemporary diaspora of "ethnic minority groups of migrant origins residing and acting in host countries but maintaining strong sentimental and material links with their countries of origin—their homelands" (Sheffer, 1986, 3).

As Castles and Miller (2009) suggest, the extent to which migrants engage in transnational behavior is difficult to assess, and there needs to be more empirical investigations into the ways that ethnic migrants negotiate their lives and maintain relations in both their country of origin and their destination society. Transnational migration also raises policy-related questions, particularly around the concept of citizenship. Theorists of transnational migration argue that migrants are likely to maintain multiple loyalties and, if possible, multiple citizenships (Powers, 2013). How will national governments react to the possibility of multiple citizenships? Some countries accept dual citizenship, others do not. How are citizenship rights to be translated in circumstances where there might be ties to multiple nations? Critics of multiculturalism are concerned about ethnic migrant loyalty and the impact of ethnic migration on national identity. Ethnic minorities moving freely between nations and maintaining strong ties to their country of origin is frightening to some people.

MIGRATION AND SECURITY

During and prior to World War II, nations such as the United States, Italy, Denmark, and Belgium were concerned that German migrants were potentially more loyal to Germany than to their national governments. For those European countries neighboring Germany, this was a particular concern, especially in light of the fact that many ethnic Germans resided in these countries and could potentially switch loyalties (Kymlicka, 2012). Hence, German migrants and German ethnic groups were perceived as a threat to national security. In the contemporary world, the group most typically perceived as a potential aggressor and a threat to national security includes Muslim migrants and members of Muslim communities.

Beginning in the 1980s, the migration of Muslims began to be viewed as a security problem (Castles and Miller, 2009). Around the time of the Islamic Revolution in Iran in 1979, a number of key events happened to create tension between Western governments and various Arab and Muslim groups. First, in 1982, the Israeli invasion of Lebanon created tensions. The Israeli invasion targeted the Palestinian Liberation Organization (PLO), which was based in Beirut. This invasion led to the creation of Hezbollah. A Multinational Force was sent to Lebanon to ensure the withdrawal of the PLO from the capital. American and French militaries, as part of this Multinational Force, were the targets of a number of suicide bomb attacks and, in 1983, the American Embassy in Beirut was bombed, killing 63 people.

Second, the Soviet war in Afghanistan, beginning in 1979, further created security issues around migration. This war resulted in a U.S.-led coalition that provided weapons to the Mujahadeen—Afghani and some non-Afghan Muslims who fought the Soviet army. The Mujahadeen is considered to be the genesis of Al-Qaida (Castles and Miller, 2009, 215). After the Soviet withdrawal from Afghanistan in 1989, and later a civil war in Afghanistan, the Taliban emerged as the primary Islamic fundamentalist political group to rule the country. Osama bin Laden and Al-Qaida began to form a significant base of operations in Afghanistan under the Taliban regime.

The first large-scale terrorist attack to reach the United States was the 1993 bombing of the World Trade Center. Arab migrants were involved in this attack. Another major attack on the World Trade Center occurred on September 11, 2001 (9/11) and led to the collapse of the Twin Towers and the death of over 3,000 people. On September 11, 2001, the Pentagon was also attacked, and 40 passengers on a hijacked plane died when the plane, thought to be on its way to the U.S. Capitol building in Washington, DC, crashed in Shanksville, Pennsylvania (Graham, 2004). Since this time, there have been a number of terrorist attacks in the United States and Europe. There was the bombing of commuter trains in Madrid, Spain on March 11, 2004 that killed 191 people, and in London, on July 7, 2005, there was a series of bombs that targeted the public transit system during rush hour, killing 52 civilians. One of the more recent terrorist attacks occurred on April 15, 2013 when two bombs exploded at the Boston Marathon in the U.S., killing three people.

Islamic fundamentalists are thought to have conducted each of these terrorist attacks in the U.S. and Europe, prompting concern about the emergence of a radical Muslim threat to security in both continents. The wars in Iraq and Afghanistan are thought to have fueled anger among some Muslims and some have volunteered to fight against the U.S.-led invasion of Iraq (Castles and Miller, 2009). According to the *New York Times* reporter Philip Shenon (2008), who investigated the Warren Commission's report on the 9/11 terrorist attacks, thousands of Muslims received military training in various camps across the Middle East and North Africa. However, it should be noted that the vast majority of migrant Muslims in Europe and the United States do not support Al-Qaida-style radicalism (Castles and Miller, 2009, 216). Also, studies have suggested that many Muslim migrant communities are assimilating. A study by Michele Tribalat (1995) found that many Arab migrant households in France speak French, and there has been a decline in traditional arranged marriages and a rise in intermarriage between migrant Arabs and French citizens. His work suggests an overall incorporation and assimilation of Arabs within French society.

In many respects, the concern over terrorist attacks and terrorist military training by Muslims has encouraged the recent "securitization" of migration (Bourbeau, 2011). Controlling migration is now seen as imperative to national security and, as we explore in Chapter 5, there is an increasing emphasis on policing national borders and migrant communities. According to Philip Bourbeau (2011), the process of tightening migration control for security reasons was in motion even before the 9/11 terrorist attacks, although he admits that government reaction to these attacks has certainly heightened the linkages between migration and security. Geographer Mathew Coleman (2007) demonstrates how, in the post-9/11 era, border policing has moved to interior spaces within the U.S., and this interior enforcement has localized migration policies in ways that attempt to control undocumented labor migration. The notion of migrants as a threat to national security has moved beyond the Muslim migrant to include the undocumented labor migrant who illegally crossed national borders.

MIGRATION AND GENDER

As we mentioned in Chapter 1, the early geographer Ernest George Ravenstein, following his examination of migration in mid-nineteenth-century Britain, postulated a number of laws related to migration. One of these laws states that women tend to migrate more than men, although, according to Ravenstein, when women migrate, they tend to move short distances. During Ravenstein's time, there were a significant number of women who migrated to neighboring counties in Britain to work as domestic servants for wealthy families and for families of the emerging middle classes of the industrial age.

What is interesting about Ravenstein's law is that it contradicts the general perception that males migrate more than women and are, somehow, innately more mobile (Gabaccia, 2013). In many studies of migration, it is the male migrant who is portrayed as the risk-taker, leaving his country for another while the female follows later. But, increasingly, this is not the case (Gabaccia, 2013). The proportion of females who migrate for economic reasons is on the rise in recent times, and in certain regions of the world, female migration is more likely than male migration. Nearly half of immigrants in the world in 2005 were female. In regions such as North America and Europe, most immigrants in that year were female. In all regions, the proportion of female immigrants increased between 1960 and 2005, and, in some cases, there was a substantial increase. For instance, in central and eastern Europe, the proportion of female migrants increased from just about half of all migrants in 1960 to 56 percent in 2005. These levels of increase have led to the use of the term, "the feminization of migration."

Yet, it should be noted that while female migration has increased in recent decades, even as far back as 1960, women made up 47 percent of all migrants. In more developed regions such as North America, during the 1960s half of all migrants were female. In Asia and Africa, female migration is less common than male migration. In the case of Africa, migrants are largely recruited by the oil industry, which demands male rather than female workers. In Asia, female migration is lower than male migration, although in some countries most migrants are female. For instance, in the case of the Philippines, 65 percent of those who out-migrated in 2005 were women. Between 2000 and 2003, 79 percent of Indonesian migrants were women. Many of these Filipino and Indonesian female migrants became domestic workers, sometimes in Persian Gulf oil states.

As we explore in Chapter 5, in the case of the United States, migration policy has tended to favor family reunion, encouraging women to join their husbands who had migrated earlier for work. This would help explain the high proportion of female migrants in North America in 1960. However, more recent female migration is arguably different. More women now migrate independently as labor migrants rather than as family dependents of male migrants. Some of these women are international students who remain in their host nation after completing their studies. Some are professionals. Still others are less-skilled workers. There is also an increasing number of women trafficked into the sex industry, and about half of all refugees and internally displaced populations are women.

Migrant men and women hold different types of jobs. At the lower end of the skills range, women tend to hold positions in the so-called light industries (textiles, garment-making, food-processing) or in domestic service. In contrast, men tend to work in mining and heavy industry (Gabaccia, 2013). Women migrating from less developed to more developed regions of the world for employment often provide daycare and cleaning services and, in some cases, eldercare. The increasing number of native women entering the workforce in developed regions has led to an increase in the need for child carers and other domestic service positions, jobs largely filled by migrant female workers from less developed countries (Gabaccia, 2013). In more high-skilled positions, women have a variety of jobs: they work in universities conducting research and teaching; they help run multinational corporations; and a large proportion of women work in the healthcare industry as nurses and physical therapists (Martin, 2004). With increasing access to higher education for women, many women enter universities in Europe and the United States as international students.

Employers often incorrectly assume that female migrants are not the sole breadwinners, and there is an expectation that women will leave their jobs when they get married. Women migrants suffer from occupational gender segregation. Occupational gender segregation occurs when certain occupations are composed of a single gender. Female migrants are often employed in low-paying jobs in the health sector, in domestic services, or in the garment and textile industries. Among female labor migrants, those in lower-skilled employment are particularly vulnerable to exploitation and can be poorly paid. According to a study of migration in Europe by Ronald Ayres and Tamsin Barber (2006, 30):

> the majority of third-country nationals are employed in vulnerable, low-skilled, low-paid jobs ... There are sharp differences in pay between men and women, which can be explained in part by women's disproportionate representation in low pay sectors, such as cleaning and

> domestic work, the casual or part-time nature of many female jobs and their concentration in the informal sector.

Female irregular migrants can be found in a variety of industries and some of these undocumented migrants are smuggled by traffickers. In some instances, women who seek the help of smugglers to enter a country find themselves deceived and subsequently trapped into forced prostitution, marriage, domestic work, or other forms of exploitation. In Chapter 5, we discuss in detail human trafficking. Women are especially vulnerable—the overwhelming majority of victims of human trafficking are women and girls. Many women are forced into the sex trade. They are subjected to mental and physical abuse, and are forced to have sex with multiple partners (often without contraceptives). According to Castles and Miller (2009, 237), 120,000 to 170,000 persons are trafficked in the European Union each year, and of those 75 to 80 percent are involved in the sex industry. Some women are exploited as domestic servants, working under very harsh conditions and, in some cases, subjected to physical and sexual abuse. Human trafficking (see Chapter 5) and the particular vulnerability of women as victims are becoming an important international policy concern.

Another area of concern is the needs of refugee women. Women are generally more likely than men to be refugees, and they can be especially vulnerable. A refugee is defined by the international community as a person who is fleeing from persecution on the basis of race, religion, nationality, political opinion, or membership of a social group (see Chapter 5). Gender is not one of the defining characteristics, and yet women often flee their country of origin because of gender-based forms of persecution including rape, domestic violence, forced marriage, female genital mutilation, widow burnings, or other forms of honor killings (Martin, 2004). Recognizing the difficulties faced by refugee women, the United Nations High Commissioner for Refugees (UNHCR) has developed certain "Women at Risk" programs, and countries that accept refugees for resettlement purposes acknowledge the needs of women refugees who are vulnerable to such gender-based forms of persecution. In some instances, women who resettle or live in refugee camps for long periods of time become more empowered against the discrimination and exploitation they faced in their countries of origin (Martin, 2004).

This can be the case of women more broadly. For women, migration can mean entry into the labor force for the first time, providing these women with economic independence and an opportunity for social mobility not previously possible. Migration by women can alter family dynamics and the relationships between men and women within a family unit. The distribution of power within the family can be altered as women achieve economic independence.

Migration theory related to new economics of migration suggests that families and communities together make decisions regarding the decision to move. Whether or not a female will migrate depends on the age of the woman, her position within the family unit, whether or not she has children, as well as the cultural norms and values of the society she is born into (Boyd, 1989; De Jong, 2002). In certain communities and societies, women are discouraged, and in some cases, forbidden to migrate.

National policies are typically not gender neutral. The migration laws of receiving countries are important in this regard. These laws can classify women as "dependents" of male migrants (Boyd and Grieco, 2003). The family role of the woman is emphasized rather than her role in the labor market. In some cases, a woman will be allowed to reside in the receiving nation but will not receive a work permit to enable her to become employed. This creates situations in which women enter the informal economy where jobs are poorly paid and low-skilled. There is a need to develop a better understanding of the role of women in the migration process, and develop policies that are more reflective of women's concerns and inclusive of women's rights.

MIGRATION AND SEXUALITY

People migrate for a variety of reasons. There is a tendency to focus on migration for the purposes of employment or better wages, but people also often migrate to reunite or remain with their loved ones. In a recent article on migration, Nicola Mai and Russell King (2009) call for a "sexual" and an "emotional" turn in migration studies, suggesting the need to consider sexuality and love as important elements of the migration process. They state (2009, 296),

> Love, whether it is for a partner, lover or friend, or for a child, parents or other kin, is so often a key factor in the desire and the decision to move to a place where one's feelings, ambitions and expectations—emotional, sexual, political, economic, hedonistic etc.—can be lived more fully and freely. Sexuality is another increasingly relevant and recognized axis of self-identification for people, whether they decide to stay put or to migrate. Like love, and sometimes alongside it, sex can play a decisive role in the imagination and enactment of the choice to migrate.

Sexual identity and feelings for another person can be strong motivators for migration. Lesbian, gay, bisexual, and transgender (LGBT) individuals can migrate in search of places that offer more freedom for them to express their sexual identity. They can migrate out of fear of persecution because of their sexuality. Sometimes they migrate to be with a same-sex partner. In these cases, LGBT people face specific challenges and obstacles to migration.

In early migration policy, homosexuality was not always explicitly referenced as a characteristic of migration exclusion, but as Audrey Yue (2013, 1) suggests, in most developed nations, "immigration policy has historically constructed heteronormative sexual norms and identities" that negatively impact LGBTs. The 1917 United States Immigration Act introduced a provision excluding persons with "abnormal sexual instincts" from entry into the U.S., and, although the term homosexuality was not explicitly used in the legislation, the understanding was that LGBT individuals would be included in this category. In 1952, the United States Immigration and Nationality Act contained within it explicit references to homosexuality as "sexually deviant," and homosexuals were described as being "afflicted with psychopathic personality" (Canaday, 2009, 214) and therefore they were denied entry. In earlier times, homosexuals were deported from the United States because of their sexuality (Coleman, 2008). Historically, LGBT individuals have been directly or indirectly excluded by migration policy, and it is only in more recent times that policies have emerged in a few countries to enable same-sex migration.

Australia was one of the first countries to accept same-sex relationships as a basis for migration. This policy was first introduced in 1985. In 1991, Australia introduced interdependency visas for foreign nationals in a relationship with a citizen or permanent resident.

Heterosexual and homosexual unmarried couples could both avail themselves of these visas (Yue, 2013). The category of interdependency was included in Australia's family migration program, and according to Audrey Yue (2008), between 1991 and 2008 some 7,500 permits were issued via the interdependency program. Audrey Yue (2013, 2) states that by 2009, 21 other countries recognized same-sex relationships as the basis for legal sexual migration. In the United States, the same-sex provision in migration policy is challenged by questions around same-sex marriage. Until 2013, same-sex marriage was not recognized at the federal level of government, based on the 1996 Defense of Marriage Act. Thus, migration policy did not permit same-sex partners and their children to be reunited as a legitimate recognized family unit. However, the U.S. Supreme Court recently ruled the Defense of Marriage Act unconstitutional. While the laws and regulations still need to be adjusted accordingly, this will most certainly affect migration policy for LGBT married couples.

Sometimes LGBT people migrate as asylum seekers. An asylum seeker is a person who has applied for international protection out of fear of persecution (see Chapter 1). Similar to gender, fear of persecution for reasons of sexual orientation is not classified as one of the defining characteristics for obtaining refugee status. LGBT people seek refugee protection on the basis of their membership of a particular social group that can be subject to persecution. In 1991, Canada was one of the first countries to accept LGBT refugees. Australia and the United States did the same in 1994. According to Audrey Yue (2013), up until 2008, only 19 countries had allowed people to receive asylum on the basis of their sexual orientation. In some cases, asylum for LGBT people has been refused because, as the argument holds, these individuals can safely return to their country of origin if they hide their sexual orientation. In other words, it is argued that they will not be persecuted if they do not reveal their sexual identity. In July 2010, this argument was rejected by the Supreme Court in the United Kingdom as inconsistent with the Refugee Convention (Yue, 2013). As countries continue to persecute on the basis of sexual orientation and sexual identity, it will be increasingly important to allow LGBT people to receive asylum in countries that are more open and accepting of their sexuality.

CHAPTER SUMMARY

In this chapter, we examined the relationship between migration and society. Migration leads to the development of ethnic communities in receiving nations. These nations can respond in different ways. Some encourage migrant assimilation in which ethnic migrants become incorporated into the society. When this occurs, they adopt the customs and ways of their host nation. We examined different types of assimilation including socio-economic and spatial assimilation.

In more recent times, nations have adopted a multicultural approach that focuses on the equal rights of ethnic migrant groups. This approach does not require ethnic minorities to adopt the ways of the dominant ethnic group but rather celebrates ethnic differences. In recent years, there has been a backlash against multiculturalism, although according to work by Banting and Kymlicka (2006), many developed nations continue to introduce multicultural policies. Concern about the multicultural approach centers on the fear that migrants will not adopt the national identity of their host nation. Recent terrorist attacks, along with increasing concern about radical Islamic fundamentalism, have heightened links between migration and security, particularly in a post-9/11 world.

We discussed gender and migration. Migration has always included the movement of women but, in more recent years, women are migrating for economic reasons. Female migrants often face discrimination in the labor market and find themselves in poorly paid gender-segregated occupations. Women are more likely than men to be trafficked, forced into prostitution, domestic work, or marriage. In addition, women are forced to migrate out of fear of different forms of persecution including rape, domestic violence, and sexual abuse. The global policy community is increasingly aware of problems associated with the trafficking and abuse of female forced migration. Better policies will result when we understand more about the process of migration in a gendered world.

Finally, we examined the policy environment for LGBT people. Some countries are molding their migration policy environment in recognition of same-sex couples and their families. Other countries, including the United States, have yet to fully adopt adequate same-sex migration policies, although this recently changed as a result of the U.S. Supreme Court's decision on the Defense of Marriage Act, opening up the possibility that married same-sex couples will receive the same federal benefits as heterosexual couples, including immigration benefits. Other countries still persecute individuals on the basis of their sexual orientation and sexual identity. Recognizing that people migrate for reasons of love and sexuality, it is important to consider the specific policy needs of the LGBT community.

GUIDE TO FURTHER READING

For an interesting book related to the topic of women and migration, see Parreñas, R. (2001). *Servants of Globalization: Women, Migration, and Domestic Work.* Stanford, CA: Stanford University Press.

For an interesting book related to the topic of migration and sexuality, see Luibheid, E. and Cantu, L. (Eds) (2005). *Queer Migrations: Sexuality, U.S. Citizenship and Border Crossings.* Minneapolis, MN: University of Minnesota Press.

To further examine migration and security, see Bourbeau, P. (2011). *The Securitization of Migration: A Study of Movement and Order.* Oxford, UK: Routledge.

To learn more about the "Multiculturalism Policy Index," see Queens University, http://www.queensu.ca/mcp/.

4

MIGRANTS AND THE GLOBAL ECONOMY

Migration greatly impacts the economy of nations, cities, and communities. In this chapter, we examine four processes important to understanding the relationship between migration and the global economy. First, we consider the relationship between migration and the labor market of host nations. In particular, we examine the impacts of migration on the workers of receiving countries. Second, we examine the concepts of ethnic enclaves and ethnic niches. Migrants tend to cluster and reside in specific neighborhoods. Ethnic businesses also co-locate in these neighborhoods. An ethnic enclave is a place where there is spatial concentration of ethnic businesses where co-ethnic residents are employed (Chan, 2013). Ethnic niches refer to particular types of businesses that are disproportionately owned and/or operated by particular ethnic groups (Zhou, 2013). In this section of the chapter, we explore the role of ethnic enclaves and ethnic niches in the economic assimilation of migrants.

Third, as we showed in Chapter 1, migrants often move from rural to urban areas. The urbanization process is, in part, the result of a rural-to-urban migration. Migrants tend to move to certain cities, changing them in the process. Typically, these are cities that are important nodes in the global economy, referred to in the literature as global or world cities. We consider migration to these types of cities, and we also examine the migration of high skilled workers and the potential impact high-skilled migration has on developing nations. Finally, migrants who move for economic reasons often send part of their earnings back home to family, friends, and their community. These remittances are thought to enhance the economy of receiving societies. We will examine the nature and effect of remittances for less developed nations.

MIGRATION AND LABOR MARKETS

In their book, *Global Migration and the World Economy*, Timothy Hatton and Jeffrey Williamson (2005, 293) review a large body of empirical studies on the effects of migration on wages and employment, focused primarily on the United States, and demonstrate that, in fact, the results are highly mixed and variable. Some studies suggest that because of migration the wages of the native-born population are reduced, although not dramatically. In other studies, wages are largely unaffected. In many studies, the effect of in-migration on domestic employment is negligible. There is much debate in the economics literature over these different findings. The classical economic model of labor markets suggests that an inflow of immigrants should lower the wages of native workers and create unemployment. The results from empirical studies are often inconsistent with economic theory.

Hatton and Williamson (2005) suggest three reasons why this might be the case. First, immigrants generally locate in specific areas where wages are high and on the rise, and unemployment is low. Therefore, immigrant inflow has a negligible effect on these economic conditions or the wage trajectory. Second, labor markets are well integrated within developed nations such that, when immigrants enter a local labor market, native workers and previous immigrants can respond by moving to another region for work. The local labor market effectively absorbs new immigrants as a result. Third, the market for goods is similarly well integrated within nations so that the boom in goods produced by new immigrants within one region is exported quickly to other regions of the country. The region with new immigrants can easily absorb the additional pressure on the regional labor market.

Yet, Hatton and Williamson (2005) suggest that the effects of immigration on wages and unemployment might be different for workers with different skill levels. In other words, some workers might be more impacted than others. In different model estimations, they suggest that, under certain conditions, the earnings of skilled workers might actually increase while the earnings of unskilled workers decrease (2005, 317). Others have found similar results. George Borjas (2004) conducted a study across different groups of workers and found that immigration is harmful to the employment and earnings of workers who are in the same skill group as migrants. Specifically, he found (2004) that if the supply of workers in a particular skill group increases by 10 percent because of immigration, the wages of native workers in that same skill group are lowered by 3 to 4 percent and the number of days worked by these native workers is reduced by two to three weeks. In his book *Heaven's Door: Immigrant Policy and the American Economy*, Borjas (2001) found, among other issues, that immigration had a particularly negative effect on the earnings and employment of minorities, especially poorly skilled black workers.

In 1990, the United States Congress appointed a Commission on Immigration Reform to recommend changes to immigrant policies at the time. As part of that work, the commission requested the National Research Council (NRC) bring together a panel of experts to understand the impact of immigration on the U.S. economy. In 1997, the NRC produced its report entitled *The New Americans: Economic, Demographic and Fiscal Effects of Immigration* (Smith and Edmonston, 1997). The report suggests that, overall, immigration had a positive effect on the U.S. economy, producing "net economic gains for domestic residents" (Smith and Edmonston, 1997, 4). The panel report suggests that immigrants are paid less than the value of the goods and services they produce, and as a result, they provide net gains to the domestic economy. Other studies have found similar benefits to the receiving economy. For example, a study by the British Home Office suggests that immigrants create new businesses and jobs as well as fill labor market gaps (Glover, 2001).

Yet, the NRC panel report also suggests that, despite overall gains, there are winners and losers in this scenario. More specifically, the panel report states (Smith and Edmonston, 1997, 5):

> Along with immigrants themselves, the gainers are the owners of productive factors that are complementary with the labor of immigrants—that is, domestic, higher-skilled workers, and perhaps owners of capital—whose incomes will rise. Those who buy goods and services produced by immigrant labor also benefit. The losers may be the less skilled domestic workers who compete with immigrants and whose wages will fall.

In other words, employers benefit from immigration. So do high-skilled workers who, in most cases, do not compete with immigrant workers. Consumers benefit because migrants produce goods and services, and these goods and services are typically at cheaper prices when produced by immigrant rather than domestic workers. The losers, on the other hand, are low-skilled workers who are competing with immigrants for jobs. The wages of low-skilled workers will fall because of the potential for hiring immigrant workers at a cheaper rate, although the panel report suggests that the decline in these wages is rather small.

The fact that immigration might have a negative effect on the wages of poorly skilled native workers does not deter encouragement of labor migration. This is, in part, because the overall economic gains from immigration are positive, and, more important, because the gains to capital are substantive. Business-owners benefit from the labor of migrants, especially the labor of poorly skilled workers. A Marxist approach to an understanding of labor migration highlights these gains and suggests that the accumulation of capital can be directly linked to the exploitation of migrant labor in the form of relatively cheap wages, which are lower than those expected by the native worker (Vogel, 2013).

According to Borjas (2001), there has been a decline, relative to the native population, in the average skills of immigrants entering the United States over the past four or more decades. Those who entered in the 1980s and 1990s were particularly less skilled. At the same time, with the rise of neoliberalism as an economic and political agenda in recent decades, there has been deregulation in the labor markets of many advanced economies. Between 1945 and 1970, industrial expansion was marked by social welfare provision and strong trade union relations that protected workers and ensured they received adequate wages and benefits. In the current era of neoliberalism, the labor market is far less regulated and the rollback of welfare has made workers, particularly low-skilled migrant workers, more vulnerable to exploitation. Neoliberal policies have encouraged the growth in temporary and contract labor, often filled by labor migrants. The informal economy within advanced economies has expanded in recent decades (Castles and Miller, 2009). Irregular migration has contributed to the growth in this informal economy, and growth in the informal economy attracts irregular migrants.

The demand for low-skilled workers among developed nations has not dissipated. Europe is managing to meet this demand by mobility within Europe, the recruitment of temporary workers, and, as Stephen Castles and Mark Miller (2009, 233) suggest, "a tacit acceptance of undocumented migration." In the case of the United States, the demand for unskilled labor is primarily met by irregular migration, and in the case of advanced economies of the Arab states of the Persian Gulf, by large-scale contract labor. Migrant workers, as Castles and Miller (2009, 233) suggest, "meet special types of labour demand, and often experience social and economic disadvantage." Yet, there are specific ways that migrants assimilate into the larger economy. This brings us to the next section related to ethnic solidarity and the role of ethnic networks in the labor market.

ETHNIC ENCLAVES AND ETHNIC NICHES

In their pioneering article in the *American Journal of Sociology*, Franklin Wilson and Alejandro Portes (1980) first used the term ethnic enclave in their exploration of the labor market experiences of Cuban immigrants in Miami. Examining survey data from 1973 to 1976, they found that many of the newly arrived Cuban refugees who worked for co-ethnic business-owners were doing better than those who worked for similar white-owned firms. According to the traditional view of economic assimilation, gradually over time immigrants progress from the secondary labor market—low-paying and high turnover employment—to the primary labor market—high-wage careers with good benefits. The research by Alejandro Portes and Franklin Wilson (Wilson and Portes, 1980) challenges this traditional view by suggesting that immigrants can gain higher than typical returns on their human capital and progress economically by working in the ethnic enclave economy.

One means by which immigrants can progress in the enclave economy is by developing new businesses. In later work, Alejandro Portes and Robert Bach (1985) found that the proportion of Cuban refugees who ran their own businesses increased from 8 percent to 21 percent from 1973 to 1979, and an important predictor of self-employment among these Cuban refugees was employment by a co-ethnic business three years earlier. By working for a co-ethnic business-owner, immigrants learn from them in ways that allow them to develop their own businesses. As in earlier research, Portes and Bach (1985) found that Cubans who worked in a co-ethnic firm earned better wages than Cuban immigrants with similar experience and education who worked in the broader secondary labor market. Even though the ethnic enclave is made up of industries in the secondary sector, immigrants working in the enclave economy receive similar benefits and wages to those in the primary sector (Waldinger, 1993).

The work by Alejandro Portes and his colleagues generated a lot of discussion among scholars around the benefits of the ethnic enclave. For instance, Jimy Sanders and Victor Nee (1987) suggested that it is ethnic business-owners not ethnic workers who gain from the ethnic enclave economy. In their work, they determined that ethnic business-owners exploit the vulnerable immigrant co-ethnic workers in the ethnic enclave and that is how these business-owners gain an advantage. A series of journal articles on the topic ensued (Sanders and Nee, 1987; Portes and Jensen, 1987). Portes and his colleagues suggested that Sanders and Nee did not correctly define the enclave since they assumed the enclave is a place of residence.

In fact, researchers define the ethnic enclave in different ways. For example, in her examination of Chinatown in New York, Min Zhou (1992) defined the ethnic enclave in three ways: as a place of residence, a place of work, and by industry. Despite its different defining characteristics, spatial concentration of some sort is typically central to the concept. Because of the need for proximity to ethnic resources such as a co-ethnic labor supply, other forms of ethnic support, and, in addition, proximity to co-ethnic consumers, ethnic businesses as well as co-ethnic immigrants working for them gain the benefits of spatial concentration. As Alejandro Portes and Robert Manning (1986, 330) suggest, "Once an enclave has fully developed, it is possible for a newcomer to live his life entirely within the confines of the community. … This institutional completeness is what enables new immigrants to move ahead." Also of importance in describing the ethnic enclave economy is the notion of the ethnic entrepreneur. These are largely ethnic business-owners who operate their businesses in an urban neighborhood where other co-ethnics are concentrated. They are very much entwined and engaged in the social network of the ethnic community.

ETHNIC ENCLAVES

Ethnic enclaves are quite common in big cities throughout the world. Migrants have generally settled in urban areas and clustered in neighborhoods with people that share their characteristics (e.g. race, ethnicity, and socio-economic status). As the largest immigrant destinations, many big cities in the United States have numerous neighborhoods, or ethnic enclaves, where migrants have settled. While New York City offers the greatest diversity of ethnic enclaves, these places are commonplace across the country in large and small cities. Historically, the economic opportunities that ethnic enclaves create are abundant: Koreatown offers grocery stores; Irish neighborhoods offer bars; Little Italy offers restaurants; Chinatown offers shopping opportunities. There are other well-known enclaves, which include Little Manila in Daly City, California; Little Havana in Miami, Florida; Little Italy in Boston's North End; and Latino barrios in Chicago—to name just a few. Indeed, there are similar enclaves located across the globe.

Outside of the central city, ethnic enclaves have also emerged in the suburbs. "Ethnoburbs," or ethnic enclaves in the suburbs, emerged at the brink of the twenty-first century. In 1998, Wei Li coined the term in her study of suburban Los Angeles, and she later refined the concept in the book *Ethnoburb: The New Ethnic Community in Urban America* (Li, 2009). Examples include Chinese migrants in San Gabriel Valley, California; Korean migrants in suburban Maryland; Latino migrants on Long Island, New York; Brazilian migrants in Framingham, Massachusetts; and Arab migrants in Dearborn, Michigan. Just as earlier migrants arrived in big cities, twenty-first century migrants have arrived and settled directly in the suburbs.

There are cultural dimensions to the ethnic enclave economy. Ethnic solidarity is viewed as important. Wrapped up in the concept of the enclave economy is the notion that relationships between co-ethnic business-owners, their clients or customers, and co-ethnic workers, in the words of Min Zhou (2013, 3) "generally transcend a contractual monetary bond." As Alejandro Portes and Robert Bach (1985, 343) state in their book, *Latin Journey*,

> Ethnic ties suffuse an otherwise "bare" class relationship with a sense of collective purpose. ... If employers can profit from the willing self-exploitation of fellow immigrants, they are also obliged to reserve for them. ... supervisory positions. ... to train them. ... and to support their. ... move into self-employment.

Solidarity and support among ethnic populations ensure that immigrants can succeed economically despite challenges of entering a new society.

When considering issues of ethnic solidarity in the labor market, research has broadened to focus less on a physical concentration of ethnic businesses and more on the concentration of ethnic groups in certain industries or occupations. The term ethnic niche is used to describe the overrepresentation of ethnic minorities in particular industries or occupations. As with ethnic enclaves, there are different ways to define ethnic niches. Roger Waldinger and Mehdi Bozorgmehr (1996) suggest that an ethnic niche exists when an ethnic group makes up 50 percent or more of those involved in a certain industry or occupation. John Logan and his colleagues (2000) define an ethnic niche as simply an overrepresentation of ethnic minorities in particular employment sectors. Ethnic niches can offer opportunities for ethnic minorities to utilize ethnic networks to gain information on potential job openings, access job training, and overcome potential barriers of racial discrimination and employment competition (Waldinger, 1996). As Roger Waldinger (1996, 303) states,

> Niching is pervasive and persistent because it is self-reproducing. Outsiders get excluded when employers rely on ethnic networks: insider's contacts are the first to find out about job opportunities; and employers know more about the associates referred by their own

employees than they do about newcomers with no prior connections to the workplace.

Some ethnic niches are advantageous. For instance, African-American concentration in the civil service has enabled many to obtain secure employment in this sector. However, ethnic niches can be damaging, limiting immigrants and ethnic minorities to specific types of jobs (Waldinger, 1996). In a series of articles in the *Journal of Ethnic and Migration Studies* in 2007, it is shown that ethnic niches are often gendered, particularly in the area of domestic work. Ethnic women are overrepresented in the domestic service industry; in part, this is a reflection of economic circumstances as well as a reflection of an ethnic niche (Moya, 2007).

Ethnic niches and ethnic enclaves provide new insights into migrant assimilation into the economy and labor market. Some have suggested that the study of ethnic enclaves and ethnic niches needs to broaden, to examine and determine the influence of globalization on ethnic businesses. In their recent book, *Chinese Ethnic Businesses*, Eric Fong and Chiu Luk (2006) explore the connections between the international flow of capital as well as how global labor migration impacts ethnic businesses and ethnic enclaves. For instance, an urban neighborhood such as Chinatown in New York city is not only comprised of small family-owned businesses but also includes foreign banks and professional companies with financial backing from Hong Kong and Taiwan. Chinese ethnic media conglomerates have emerged in the United States as a result of increased financial investment and growth in the Chinese community because of migration (Zhou and Cai, 2007). The Chinese banking sector has grown in California because of availability of transnational capital and the increase in Chinese wealth (Li and Dymski, 2007). Many ethnic businesses have transnational connections that can be beneficial. Products, labor, and consumers for ethnic businesses can more easily cross national boundaries as the global economy becomes more integrated.

GLOBAL CITIES AND HIGH-SKILLED MIGRATION

Scholars of globalization have drawn attention to the ties between economies of such cities as New York, London, Tokyo, and Los

Angeles and global economic conditions (Erie, 2004; Sassen, 1991; Scott and Soja, 1996). These and other cities are major international financial centers and nodes of coordination for economic activities for the world's capitalist markets, and have received much attention in the literature on globalization. Global cities research was initially limited to just a few large cities, such as New York, London, Tokyo, and Los Angeles. It has since expanded to include many more large cities, using many different ways to identify the relationship between globalization and the city.

For example, scholars from the Globalization and World Cities (GaWC) research group (Taylor, 2004) use quantitative methods to identify global connectivity based on the distribution of advanced producer services among a range of cities. Those cities that have at least one-fifth of the connectivity of the most connected city, which was London, are identified as world cities. These include cities such as New York and Tokyo but also include cities such as Madrid, Istanbul, Dubai, São Paulo, and many others. As Marie Price and Lisa Benton-Short (2008) point out, *Foreign Policy*'s annual globalization index considers four criteria to identify global connectivity including economic integration, political integration, technology such as the Internet, and personal contact. So while the focus is on economic integration, particularly around financial services, there are many ways cities connect and are integrated within the global society.

One important way that cities become defined as global in nature is through migration. John Friedmann (1986, 75), in his initial formulation of the world-city hypothesis, states "world cities are points of destination for large numbers of domestic and/or international migrants." Migration is significant in classical cities of immigration including the traditional global cities of New York, London, and Sydney but, as Marie Price and Lisa Benton-Short explain, migration is a defining element of many cities. Price and Benton-Short (2008, 6) classify these cities as immigrant gateways, observing that they are "nodes for the collection and dispersion of goods, capital and people." As nodes emerge, re-emerge, or remain central to the global economy, they attract migrants. In their book, *Migrants to the Metropolis*, they identify emerging immigrant gateway cities including Singapore, Washington, DC, Dublin, and Johannesburg; exceptional immigrant gateways including cities such as Tel Aviv,

Seoul, São Paulo, and Riyadh; and the established immigrant gateways of New York, Sydney, Toronto, and Amsterdam. Different migrants from different parts of the world are living, working, and experiencing many different types of cities across the globe.

Saskia Sassen (1988), in her classic research, demonstrates the importance of migration and flexible labor markets in the growth of large global cities. As Sassen (1988, 1991) points out, global cities, because of their importance in the global economy, attract immigrants both for high-skilled as well as low-skilled jobs. Global cities act as major centers for the global knowledge-based economy, and they need a lot of people with many different skills and abilities to function. Businesses centered in these cities are international in their perspective and activities, and they need an increasingly more mobile group of elite workers to be successful.

In global cities, the labor market is very segmented. Migrants in low-skilled service jobs work largely to support the lifestyle of the people working in knowledge and financial service jobs concentrated in these cities. These low-skilled jobs include occupations such as domestic servant, taxi-driver, janitor, gardener, childcare worker, hairdresser, waitress, security guard, and other positions that cater to well-off and busy high-skilled workers (see Chapter 6). Yet, some migrants are highly skilled in technical occupations including engineers, doctors, and information technology workers. Migrants engaged in technical and highly skilled jobs arrive in their new destination city with specific knowledge and ability, often recruited or "headhunted" by global firms.

Compared with the large numbers of low-skilled migrants arriving in large cities across the globe, the numbers of highly mobile, high-skilled migrants are much smaller. Yet, the migration of these highly skilled professionals is significant in other ways. Some of these high-skilled migrants work for transnational corporations (TNC) as members of a transnational managerial elite, moving between companies, sometimes on inter-company transfers (Beaverstock, 2005), and are important players in the circulation of particular knowledge, skills, and intelligence, which is central to the workings of the global capitalist economy. As Jonathan Beaverstock (2005, 248) suggests, "the key function of the transnational managerial elite is to support the world city's corporate economy." Many of these elites are hyper-mobile, working and living in different cities for short periods of time or living in two or more cities at once. They have international careers. The migration process is quite different for transnational managerial elites than it is for low-skilled migrants. They find it much easier to move from city to city, adapting easily to a new environment because they are supported by the policy regime (arrangements for visas and other necessary documentation taken care of by the company) and accepted by the receiving polity and society as important assets to the receiving and, indeed, the global economy (Hannerz, 1996).

Some high-skilled migrants are part of the corporate elite. Others work as doctors, lawyers, teachers, engineers, information technology specialists, teachers, university professors, scientists, and other highly skilled occupations. These workers are concentrated largely in big-city regions. They are often highly educated, the products of the universities and educational institutions of their nation of origin. The reasons why highly skilled people migrate include the potential to earn higher wages and acquire more benefits in the receiving nation than in their country of origin. In many cases, high-skilled migrants will move because they will potentially receive more international prestige if they succeed in one of the more advanced economies. In addition, other factors affecting the choice of location to where high-skilled migrants move relate more specifically to linguistic, cultural, and geographic proximity (Belot and Hatton, 2008). Many times high-skilled migrants move to nations where there are colonial connections (e.g. Indian professionals migrating to the United Kingdom) or where they can speak the language.

Private companies and state institutions are engaged in the active recruitment of skilled migrants to fulfill particular labor needs. This has become much easier with the ability to advertise employment via the Internet. Large corporations use sophisticated software programs to screen applicants and manage databases of résumés that can be accessed quickly if a job opens (Iredale, 2001). International recruitment agencies are central to companies interested in more easily probing the global labor market. As we explore in Chapter 5, many governments tailor their immigration policies to select high-skilled migrants. As Robyn Iredale (2001, 16) suggests, "the internal labour markets of transnational corporations are now largely uninhibited by the workings of government."

Certain industries are more fluid than others in terms of skills requirements and standardization and accreditation of qualifications. For example, there are many doctors from developing countries in particular who end up working as taxi-cab drivers and waiters in receiving nations because their qualifications are not recognized there. Yet, in some industries, on-the-job training and qualifications are as important as educational certifications. As Robyn Iredale (2001, 10) states, "accreditation of qualifications is increasingly driven by private employers who have established their own internal training systems (e.g. Microsoft)." This is particularly the case for information technology (IT) where there is a high level of mobility between countries and companies (Iredale, 2001).

THE MULTINATIONAL CORPORATION

Multinational corporations (MNCs) are global companies with a global labor force. As the economy globalized, large corporations entered new markets in other developed and developing countries. While there are thousands of multinational corporations, the global economy is dominated by a select group of the largest. These include banks (Barclays, UBS, Citibank), information technology corporations (Microsoft, Google) retailers (Walmart, Carrefour), and corporations involved with raw materials (Exxon, BP). These corporations establish offices in other countries where they conduct business. Just like local companies, they compete to manufacture, buy, and sell goods and services. Multinational corporations employ a labor force with both high and low skills. While the proliferation of global business has undoubtedly strengthened economies and opened new markets, it has also created tensions in other countries where MNCs locate. They can limit competition and force local businesses to close. Also, they can influence the supply and demand of the labor force, sometimes underpaying workers. Such corporations are essential to the connectivity of the global economy, and they have created local economies for new goods and services while also attracting migrants in search of new employment opportunities.

The internationalization of higher education has also been an important factor in the migration of high-skilled people (Iredale, 2001). Migrant students from all over the world attend higher education institutions in nations such as the United States, the United Kingdom, Australia, and Canada. Many times, they remain in one of these countries once they obtain their degrees. The United States admitted 565,000 foreign students in 2004 (Koser, 2007), and more than half were from Asian nations. In some cases, students in less developed countries study in their home countries but their degrees are awarded by higher education institutions of more developed countries, sometimes with an eye to the student migrating there. Australian, North American, and European universities have branch campuses in India, China, and other less developed nations. The universities of more advanced economies offer online courses and degrees that students from the developing world can access.

Migrants who undertake their training in their destination country tend to receive higher wages and are less likely to be unemployed than migrants trained in their country of origin (Coulombe and Tremblay, 2009). In some cases, migrants enter a new destination with an undergraduate degree, obtain a graduate degree in their destination country, and then remain there to work. For instance, many immigrants from India who left with an undergraduate degree in engineering from one of the Indian Institutes for Technology (IIT) went on to receive postgraduate education in the United States, to go on to fuel the high-tech industry in Silicon Valley (Khadria, 2013). As Stephen Castles and Mark Miller (2009, 65) state,

> In recent years, the USA, the UK, Canada, Australia, and Germany have all changed their immigration rules to encourage such graduates to stay—especially those in electronics, engineering and science. Indian and Chinese PhDs form the scientific backbone of Silicon Valley and other high-tech production areas.

Migration of the highly skilled can result in the loss of the most educated and accomplished in the sending society. When this movement leads to significant depletion of highly skilled and educated people within a sending nation, it can result in a process referred to as the brain drain. Sending nations lose not only the highly skilled and educated individuals as members of the workforce, but they also lose any potential return on investment that the country may have made in educating and training these migrants.

According to data analyzed by Frédéric Docquier, Lindsay Lowell, and Abdeslam Marfouk (2009), some of the largest international suppliers of high-skilled migrants in 2000 included the United Kingdom (1.5 million), India (1.03 million), the Philippines (1.1 million), Mexico (0.95 million), Germany (0.94 million), and China (0.78 million). As Khalid Koser (2007) suggests, there has long been concern that educated migrants from Europe leave for North America for higher wages and better benefits. Poorer countries are also affected. The most educated emigrants are often from developing parts of the world. For instance, a high percentage of college graduates leave places such as Haiti, Sierra Leone, Ghana, Kenya, and other less developed nations, sometimes to move to other poor nations but often to more advanced economies. Brain drain is of particular concern when the direction of the movement of high-skilled individuals is from less developed nations to those that are more economically advanced. Although it is difficult to discern the exact numbers of high-skilled migrants, Peter Stalker (2000, 107) estimates that there were about 1.5 million skilled migrants from developing countries living in the more advanced economies of western Europe, the United States, Japan, and Australia.

The loss of doctors, nurses, and other medical personnel from nations in sub-Saharan Africa and other poor parts of the world has received most attention. According to Khalid Koser (2007, 52) some of the figures are alarming. He offers some examples. For instance, only 50 out of 600 doctors trained in Zambia since this nation gained independence are actually practicing in the country. Many have gone to developed nations such as the United Kingdom. Khalid Koser (2007, 52) states, "It has been estimated that there are currently more Malawian doctors practicing in the city of Manchester in England, than in the whole of Malawi." Stephen Castles and Mark Miller (2009) state that more doctors born in countries such as Haiti, Fiji, Mozambique, Angola, Liberia, Sierra Leone, and Tanzania are working in the advanced economies of the world than in their own home countries. They also show that nine out of ten nurses from Jamaica and Haiti work in advanced economies. The impacts of these statistics are particularly devastating for poorer countries where the need for health professionals is most potent. According to Kimberly Hamilton and Jennifer Yau (2004), 37 out of 47 sub-Saharan African nations do not have the minimum recommended World Health Organization (WHO) standard of 20 doctors per 100,000 people. They also found that, in 2003, Malawi filled only 28 percent of vacancies for nursing positions and South Africa reported the need for 4,000 doctors and 32,000 nurses that same year. The movement of health professionals from developing nations has devastating impacts when the needs are so great, especially in light of the seriousness of such epidemics as HIV-AIDS and malaria in developing parts of the world.

Many argue that the movement of high-skilled migrants from developing nations to more advanced economies exacerbates global inequality and ensures that human capital remains scarce in countries where it is most needed (Docquier and Rapoport, 2011). Brain drain is particularly serious when wealthier countries actively recruit migrants with certain skills (Koser, 2007). Recently, however, governments, international development agencies, and certain scholars have become more optimistic, suggesting that high-skilled migration can bring benefits to sending nations. The concepts of brain gain or brain circulation have emerged to imply that poor, sending countries benefit when migrants return from abroad with experience and additional human capital gained during their period working in more advanced economies.

Typically, in order for brain circulation or brain gain to occur, migrants must return to their country of origin. High-skilled migrants in this case will only move to a new country temporarily. Taiwan is often cited as being extremely successful at encouraging many of its high-skilled migrants to return. In the 1970s and 1980s, Taiwan suffered classic brain drain with many college graduates migrating abroad for postgraduate education and remaining in their new destination. According to Kevin O'Neil (2003), during this time period, some 20 percent of Taiwanese graduates moved abroad and few returned. Toward the end of the 1980s and the beginning of the 1990s, the brain drain reversed and many highly educated Taiwanese, some with business experience, began to return. According to O'Neil (2003), between 1985 and 1990, some 50,000 Taiwanese migrants returned from abroad. The high-skilled return migrants were attracted by Taiwan's growing high-tech industries. Many of the large Taiwanese firms that developed during the late 1980s and into the 1990s were managed by returning migrants who had gained experience working in Silicon Valley. The Taiwanese government, through incentives, subsidies, and infrastructure developed the Hinschu Science-based Industrial Park, beginning in 1980. Imitating Silicon Valley in the United States, this industrial park enticed many U.S.-educated Taiwanese to return and start businesses. Similarly, India's information technology (IT) sector benefited from return migration as experts, educated and trained in the United States, returned to India (Khadria, 2013). Many educated Chinese have also returned as the economy of China has grown in the past 20 years or more.

During the 1990s, nations adapted policies to encourage the temporary migration of highly skilled professionals (Iredale, 2001). An example is the implementation of the H1-B visa system in the United States. This program allows companies and institutions in the United States to temporarily hire foreign workers for certain specialty occupations. Foreign workers temporarily fill positions in biotechnology, engineering, mathematics, medicine, law, accounting, information technology, and other high-skill industries. These temporary high-skilled workers return to their country of origin with experience and training.

Many of the high-skilled recipients of temporary admittance to the United States come from Europe and Asia. According to Adela Pellegrino (2001), 75 to 80 percent of those who receive visas to enter as skilled temporary workers to the United States are from Europe and Asia. Less than 10 percent of total recipients are from Latin America, which greatly impacts the ability of Latin American nations to benefit from the potential brain circulation that Asian nations have experienced in recent decades.

Temporary migration is often seen as a win-win situation where receiving nations obtain needed workers but without the social costs associated with integration, and sending nations gain through remittances and the return of skilled workers. Temporary admittance, particularly of skilled workers, encourages their return but the impact of this return, as Khalid Koser (2007) suggests, depends on the conditions that exist at home. To successfully utilize the skills and experience earned in receiving nations, return migrants need economic conditions in their home country to be such that they can obtain a job or set up a fruitful business. It is only when conditions are right that return migrants can contribute to the growth of their home economies.

MIGRATION AND REMITTANCES

Flows of international remittances to developing countries were estimated to be US$401 billion in 2012. Projections suggest that the flow of remittances to developing countries could be as much as US$515 billion by 2015 (World Bank, 2013). India, China, Mexico, the Philippines, and Nigeria received the most remittances in 2011. Smaller countries such as Tajikistan, Liberia, Kyrgyz Republic, Lesotho, Moldova, and others received the most remittances as a share of their Gross Domestic Product (GDP). The source nations for most remittances include the United States, Saudi Arabia, Switzerland, Russia, and Germany. The United States leads this group by far with US$52 billion sent out of the country as remittances in 2011.

Formal remittances are tracked through formal channels such as banking systems. Yet, remittances also occur by informal means. Migrants take home money when they return for visits or they send money home with friends and relatives (see Koser, 2007). Somalia has a very sophisticated informal system for sending money to migrant families. This is referred to as the hawilaad system. It is based on Somali traders who collect money from Somali migrants, purchase goods with this money, sell the goods in Somalia, and then pay the migrant families. The traders keep any profit made on the sale of goods. According to Koser (2007), informal remittances could be as much as twice the value of formal remittances. In that case, estimates for remittances to developing countries alone could be as much as US$802 billion for 2012.

The international flow of remittances has increased dramatically over the past decade and more. Estimates suggest that, in 1990, the inflow of remittances worldwide was US$64 billion, rising steadily to reach an estimated US$533 billion by 2012. Surveys have been conducted to understand more about who sends remittances. For example, a survey by the Inter-American Development Bank (2004) suggests that eight out of every ten Mexicans send remittances on a regular basis. According to Alex Julca (2013), undocumented migrants are more inclined to send money home to their families, perceiving this as one of the main benefits of staying in their receiving nation.

Migrants must spend money to send money home through formal channels. There are fees incurred by migrants when they send money through these formal systems. Table 4.1 indicates the average cost incurred for sending money from the United States to a selection of countries in 2004. As this table indicates, this means that for every US$100, an Egyptian migrant sends to his or her family back home, he or she must pay, on average, more than US$13 in fees.

Migrants often use money transfer companies, the two largest and most significant being Western Union and Money Gram. In a recent *New York Times* article, journalist Natalie Kitroeff (2013) wrote about the effect of changes in money transfer fees on the lives of migrant Carmen Gonzalez and her family. When Carmen sent money from the United States to her family in Mexico in 1991, it cost her US$12 for every US$100 she sent. Since then, fees to send money from the United States to Mexico have reduced. Carmen now spends US$5 to send US$100 back home. This difference is huge, and it means she can send a little more to her family. Natalie Kitroeff (2013) attributes the reduction in fees to

Table 4.1 Average cost of sending money from the United States to select nations, 2004

Country	*Percent of remittance amount*
Egypt	13.8
Venezuela	10.5
Mexico	9.2
Dominican Republic	8.4
Bolivia	8.4
Philippines	8.2
India	8.1

Source: Koser (2007).

increased competition among money transfer companies. Western Union and Money Gram are now competing with smaller companies, and, as a result, they have lowered their money transfer costs. The reduction in remittance fees for Mexican migrants is also likely the result of a new Internet platform, Calculadora de Remesas, that allows remittance senders to compare transfer fees across the different companies (Julca, 2013). According to the *New York Times* article, Western Union controlled 75 percent of the global remittance market in the late 1990s, but this has dropped to about 15 percent. Over the same time period, Money Gram's share has declined from 22 percent to 5 percent of the market (Kitroeff, 2013). Worldwide, the costs migrants incur for sending money home have declined, from about 15 percent for every US$300 in the 1990s to about 10 percent for the same amount of money in 2013 (Kitroeff, 2013).

According to Natalie Kitroeff (2013), the reduction in money transfer fees between Mexico and the United States was about US$12 billion over the 2000s. This is equivalent to five times the amount of official aid from the United States to Mexico during that same time period. Remittance fees vary considerably across regions and nations. They are generally highest among sub-Saharan African nations. In fact, South-to-South remittance fees are higher than North-to-South remittance fees (Ratha, 2013). In the South-to-South exchange, the local currency is typically converted into U.S. dollars or Euros and then converted from these currencies into the local currency of the destination nation (Ratha, 2013). This adds considerably to the cost of sending the money. International development agencies and institutions are extremely interested in figuring out ways to lower remittance fees, recognizing this as an important aspect of development policy.

At times remittances exceed official and private flows of capital. They can be a far more steady flow of funding to developing countries than official aid. Many experts, policymakers, and scholars in the field of development are excited about the potential benefits of remittances to poorer countries. Remittances can act as a form of insurance against sudden economic hardship for a migrant's family or community. Migrants often send money home when there is some sort of crisis or economic downturn (Ratha, 2013).

According to some research, remittances can also alleviate poverty in poor countries. Richard Adams and John Page (2003) conducted a study of the impact of international migration and remittances on poverty in 71 developing countries and found that a 10 percent increase in remittances leads to a 3.5 percent reduction in the proportion of people in poverty. According to Khalid Koser (2007), in Lesotho remittances are 80 percent of rural household incomes, and in Somaliland, average household incomes are doubled because of remittances. Households that receive money for healthcare and educational experiences also tend to be better off than other households. Surveying some nations, researchers have found that children of those households receiving remittances are less likely to drop out of school. They have also found that the average birth weight of children born in remittance-receiving households is higher than that of children born in non-receiving households (Ratha, 2013). Finally, remittances add currency to remittance-receiving nations, helping to pay for imports and helping these nations pay back their debts. In some instances in remittance-receiving nations, commercial banks can use the flow of remittances as collateral, improving their ability to obtain loans at lower interest rates and longer pay back times. In this way, remittances can help improve access to credit for developing nations.

On the negative side, some argue that remittances can increase the level of inequality within migrant sending communities. For example, Donald Terry and Stephen Wilson (2005) found that more than eight out of ten households receiving remittances in a village in Mexico own their own homes. Yet, only three out of ten households who do not receive remittances and live in the same village own their own home. As such, remittance households become better off than non-remittance households. Also, since migrants tend to come from particular regions within a nation, remittances can increase regional inequality (Koser, 2007).

There is concern that remittances can create a "culture of migration" (Koser, 2007). People rely on remittances and migration rather than seek to change the existing structure in migrant sending nations. Some suggest that remittances can be seen by those remaining behind as a disincentive to work. More important, migrants can face tremendous pressure to send money home, and they often have more than one job or are sometimes forced into prostitution or criminal activity in order to afford to remit. The benefits of remittances depend on how the money is spent. Remittances can be spent on consumer goods such as food, which obviously benefits the immediate family and creates demand within the economy. However, the benefits of remittances for the long-term stability of nations are greater if these monies are used for the development of infrastructure such as schools, roads, health clinics, or small businesses and community initiatives. The latter can be difficult without some coordination around how remittances are spent. The challenge for the international community is to actively facilitate remittances in ways that ensure they are useful for development purposes in developing nations.

CHAPTER SUMMARY

Economies of advanced nations benefit from the inflow of low-skilled migrants. As we explored in this chapter, because immigrants are paid less than the value of the goods and services they produce, they provide net gains to the domestic economy. Low-skilled migration can negatively impact the wages of low-skilled native workers. Yet, the overall economic gains from immigration are positive, and, more important, the gains to capital and businesses are substantive. As such migration continues to be acceptable on an economic level.

Migrants can assimilate economically when they become part of the ethnic economy. Ethnic enclaves and ethnic niches are two terms we examined in this chapter. Ethnic enclaves are places where there is spatial concentration of ethnic businesses where co-ethnic residents are employed. Some researchers suggest that migrants employed by co-ethnic businesses tend to gain higher wages and benefits than others with similar skills and education working in the broader economy. Ethnic niches include those industries and occupations where ethnic minorities are over-represented. Niches can be beneficial since they offer migrants increased access to certain jobs, training opportunities, and protection from racial discrimination. The nature of ethnic businesses is changing as the economy continues to globalize. The international flow of both capital and labor has ensured the development of different types of businesses, moving beyond the typical characterization of ethnic business as small, family-owned enterprises. It now includes larger conglomerations such as media outlets, banks, and professional services.

Migrants tend to move to large cities. Some of these migrants are highly skilled. The movement of high-skilled migrants from developing nations to developed nations is a concern. This can lead to the phenomenon referred to as brain drain. Developing nations lose the more talented and educated members of their society, which negatively impacts their ability to grow economically. In recent years, there is evidence of brain gain as high-skilled migrants return home with additional skills, experiences, and education acquired during their time abroad.

In more recent years, there has been excitement within the international development community over the potential positive effects of remittances. Migrants send money home to assist and support family and friends left behind in their home country. From a policy perspective, controlling the fees associated with sending money from immigrant receiving nations is important since it can greatly impact the amount migrants can afford to provide for their families and communities left at home.

Overall, there is tremendous interaction between migration and the global economy. Migrants generally move for employment and greater entrepreneurial opportunities. As a result, they greatly change the receiving economy and the money they earn and the skills and education they acquire can change their home nations.

GUIDE TO FURTHER READING

For additional reading on the economics of immigration, there is a volume of work that is comprised of seminal papers on this topic. See Chiswick, B.R. and Miller, P.W. (Eds) (2012). *Recent Developments in the Economics of International Migration*. Northampton, MA: Edward Elgar Publishing.

For an excellent source for data on inflows and outflows of remittances, see http://econ.worldbank.org/. For an excellent source for data on remittance prices, see http://remittanceprices.worldbank.org.

For additional reading on the brain drain and remittances, we suggest Schiff, M. and Özden, C. (Eds) (2006). *International*

Migration, Remittances and Brain Drain. New York: Palgrave Macmillan. Also, see Mohapatra, S., Ratha D., and Silwal, A. (2011). *Outlook for Remittance Flows 2011–13*. Migration and Development Brief 16. Migration and Remittance Unit. Washington, DC: The World Bank.

For additional reading related to global cities, we suggest Short, J.R. and Kim, Y.H. (1999). *Globalization and the City*. Upper Saddle River, NJ: Prentice Hall.

MIGRATION AND POLICY

In this chapter, we examine the public policy environment around labor migration and population displacement. We begin the chapter by discussing the early evolution of migration policy; how this policy is tied to the labor needs of nations and concerns about irregular migration; and how the displacement of populations because of war, the environment, and human crisis is central to global migration policymaking. As Alexander Betts (2011) reminds us, there is no "United Nations Migration Organization" with institutional or global political power to govern or to make rules to directly administer migration across international borders. No explicit institution oversees global migration. This is not to say that global institutions do not impact migration and migrants. For example, institutions such as the United Nations (UN) and the International Labor Organization (ILO) are important actors in the regulation of states' behavior toward migrants, in the case of the UN toward refugees, and in the case of the ILO toward labor migrants. But in general, national governments have been reluctant to surrender regulatory power over migration to any one international or supra-national authority.

More than international institutions, it is national governments that directly shape migration policies, albeit that nation-states continue to experience great difficulty in fully controlling migration, especially during this era of postindustrial globalization. As Stephen Castles states (2004, 852), "Observers of international migration are often struck by the failure of states to effectively manage migration and its effects on society." As we reviewed in Chapter 1, the global flow of trade, finance, and capital investment is characteristic of the contemporary economic regime. Global economic expansion coincides with the transnational mobility of people. Restrictions on the migration of people have, in general, not been relaxed in concurrence with global market liberalization (Tan, 2007). This is especially the case in the area of labor migration. Yet people continue to migrate for many reasons: to secure new employment opportunities, to reunite with family and friends, and for a better quality of life. Despite attempts by nation-states to control migration, irregular migration occurs, and the evidence suggests that it will persist throughout the twenty-first century.

In this chapter, we examine irregular migration and, in part, we consider its relationship to issues of border control and border security. The intensification of border control by different nation-states has incurred human costs as migrants take greater risks to reach destination countries. We begin the next section with a discussion about the early evolution of migration policy. Then we examine the relationship between migration policy and labor needs. We examine different policies aimed specifically at irregular migration. We discuss how migration policy occurs at different scales of political geography. Then, we examine public policies and politics regarding refugee and asylum seekers. Of great importance here is the evolution and mandate of the United Nations High Commissioner for Refugees (UNHCR) to protect and safeguard the rights of refugees worldwide. The growth of the international refugee population is a major concern for nations and global institutions, especially those concerned with human rights. In the final section of this chapter, we discuss the insidious problem of human smuggling and trafficking, a growing human rights issue. Our primary goal in this chapter is to introduce the reader to the issues related to migration policy formation and the roles of both international institutions and national governments in migrant regulation.

INSIDERS AND OUTSIDERS

Early humans began to migrate and congregate in cities about 10,000 years ago, and according to some of the early writings and pictograms, towns of several thousand people existed in Egypt and Mesopotamia up to 8,000 years ago (Harzig and Hoerder, 2009). As city formation evolved, cities grew, and even in these early times, they grew quite large. For instance, by 600 BC, the city of Babylon had some 200,000 inhabitants. People moved in and out of these early cities. Unskilled migrants moved there to build rudimentary roads, housing, dykes, and dams. Craftsmen and traders traveled to cities and sometimes settled there. The beginnings of various civilizations—including the Egyptians, the Persians, the Minoans, the Phoenicians, and later the Greeks and the Romans—all centered on the growth of cities with the inflow of migrants.

Over the years, organized and powerful elites emerged in cities. It thus became necessary to protect cities. Walls were erected, and these walls enabled an obvious distinction between those who lived inside and those who lived outside the city. As Christiane Harzig and Dirk Hoerder (2009,19) suggest, "[m]igrants, mobile farmers and merchants asked permission to enter [the city walls]." People from outside the city or territory had to be welcomed into the fold, and often wars erupted as one group moved in on another's territory to seize and take possession. As we demonstrate in this chapter, the process of migration control has long been wrought with notions of those accepted as "insiders" versus those considered "outsiders," and at certain times throughout history, migrants moved at their peril.

This is also true today. Migrants nowadays need permission to enter a nation-state and, as we explore later in this chapter, it is becoming more perilous for people to migrate across international borders either as refugees or as irregular migrants without documentation. Even in early times, it was necessary for people to have documentation to ensure safe travel or safe entry, especially during occasions of conflict. During the reign of William the Conqueror in the eleventh century, the passport emerged as a form of documentation for migratory movements. William established five control points along the coast of England where one could gain entry only with official approval. People entering England could only pass through one of these ports—hence the term "passport." Rudimentary control of migration in the form of safe-travel documents was established early on in an attempt to regulate people's movements.

Of course, passports continue to be necessary to move from nation to nation, and presenting documentation typically in some form of an international visa is required. With the evolution of the modern state, monitoring migration has become much more sophisticated. As we explore in this chapter, in contemporary times, those who migrate from one country to another without the proper documentation are branded "illegal." As Düvell (2011, 79) states, "Nowadays, the concept of 'illegal immigration' is so frequently used in public and policy discourses, and has became so common, that it tends to be forgotten that this has not always been the case." In the history of human migration, "illegal immigration" is a relatively recent branding.

In the following sections, we highlight the importance of two forces in the evolution of migration control: labor needs and displacement. As we demonstrate, in times of economic stress, the politics of migration becomes highly contentious, and, at these times, concern arises at the national level regarding the impact of labor migration on domestic wages and workers. In contrast, during times of economic need and labor shortages, migration can be actively encouraged. Sometimes migration policy is contradictory, encouraging various types of migrants while discouraging others.

In recent decades, there has been an increase in irregular migration, and we discuss some of the strategies employed to help curb this perceived problem. The ensuing debate and character of migration policy can become closely allied with notions of national identity and race. In recent years, large-scale migration has led to extraordinary diversity in many countries and communities. Sometimes this has taken place in communities that never before experienced migration. People witness demographic shifts in their neighborhoods and their nation and, feeling threatened, they tend to hold closer to their identity, culture, and language. In light of the changing nature of migration, individual states have introduced a range of measures. In addition, multilateral and supra-national regulatory systems have evolved.

In the case of displacement, global institutions and legislative frameworks have emerged to manage the dislocation of people that can occur during periods of war or unrest. Much of this evolution centers on the international concern for human rights following World War II. Policies have developed to cope with forced migration due to conflict, fear of persecution, and changes in the environment.

Let us first examine the early evolution of migration control in the face of economic considerations and labor needs.

LABOR NEEDS AND MIGRATION POLICY

For the sake of selecting a starting point in this evolution, we begin with the mercantilist period. It was during this time when the need to control migration became an important part of a nation's economic strategy (Moses, 2006).

Mercantilism as an economic system was based on the premise that national wealth could be secured by acquiring precious metals such as gold and silver. Of equal importance was the desire for strong domestic production for the purposes of exporting goods to other countries. To aid in the latter process, it became paramount to increase and stabilize the domestic labor force. Controlling the movement of workers was vital to early capitalist production. Out-migration of the labor force to competing countries was greatly discouraged and in-migration, particularly of workers with certain skills, was actively encouraged. During the mercantilist period, countries such as England, Germany, and France imported skilled labor from surrounding countries (Moses, 2006). In the 1820s, Britain banned the emigration of skilled workers as it was building its industrial base. At this time, dominant economies of Europe were reaping raw materials from their colonies and slowly populating these places. Migration to the colonies for the purposes of gaining control was actively encouraged. For instance, Australia organized publicity campaigns to recruit people from Britain, and helped to organize and subsidize their travel to the colony.

The economics and politics of migration became indelibly coupled as dominant nations of the early colonial period sought to build their economic capacity. With the Peace of Westphalia—a series of peace treaties signed to end the Thirty Years' War (1618 to 1648) in the Holy Roman Empire, and the Eighty Years' War (1568 to 1648) between Spain and the Dutch Republic—sovereignty among different states across central Europe was established. The nation became the primary political unit and controlling population movements between nations became more politically feasible. The regulation of migration emerged strongly with modern conceptions of sovereignty and nationhood combined with the rise of a more global economy.

As global industrial capitalism rose, migration control continued to be heavily tied to the economic needs of dominant nations of the West. The highly regulated guild system of the mercantilist period was discarded in favor of large-scale mass production and the ideological promotion of laissez faire markets. As we explored in Chapter 2, weakening economic barriers to migration, the active recruitment of laborers, particularly to North America, and the emergence of an industrialized global economy resulted in mass transatlantic migration between about 1820 and 1920 (Castles and Miller, 2009). Foreign workers were greatly welcomed in the United States up until the 1880s. However, during the late nineteenth and early twentieth centuries countries of the New World began to impose real restrictions on migrant entry, in many cases by specific groups of people (Hatton and Williamson, 2008). Ironically, with the rise of liberal and, later, neoliberal economic regimes, the freedom to migrate became more restricted by state control.

Arguably the first immigration law introduced in the United States was the Chinese Exclusion Act of 1882, suspending Chinese immigration and barring Chinese citizenship to the U.S. In 1885, the Alien Contract Labor law was passed, making it unlawful to contract with unskilled laborers abroad and import them from overseas, although these regulations did not pertain to crossing land borders. In the mid-nineteenth century the United States often imported groups of workers under a system of contract labor not too unlike the earlier practice of indentured servitude. The Alien Contract Labor law was largely aimed at the Chinese "coolie" system of contract laborers. The concern among legislators and labor unions was about the impact of primarily unskilled Chinese immigration on the domestic labor market. The politics of migration was heavily influenced by the potential impact of migrants on domestic wages and workers. One interesting side effect of the Chinese Exclusion Act of 1882 was the smuggling of Chinese migrants into the United States via the Mexican border. Federal law enforcement officials protecting the border were referred to as "Chinese inspectors," deployed to curb this migration of Chinese to the United States via Mexico (McDonald, 1997, 74).

Reactions to migration are bound by conceptions and attitudes about race and ethnicity. The United States Immigration Commission, known as the Dillingham Commission, was formed in 1907 in response to concerns about immigration, particularly from southern and eastern Europe. During the early 1900s, nativist groups in the United States claimed that southern and eastern Europeans could not fully assimilate into American society, and the U.S. Congress subsequently enacted the Immigration Act of 1924 designed to limit entry from countries other than those from northwestern Europe (Castles and Miller, 2009). This legislation limited the annual number of immigrants who could be admitted from any country to 2 percent of the number of people from that country who were already living in the United States in 1890.

One of the first Acts of the newly formed Federal Parliament of Australia in 1901 was the introduction of the White Australia Policy aimed at excluding Asian migrants. Southern Italians were welcomed into Australia in the 1920s, largely to work in the sugarcane plantations, replacing the blacks expelled with the introduction of the White Australia Policy 20 years earlier. In Canada, similar to the United States, the Chinese Immigration Act of 1923 restricted most forms of Chinese migration into the country until it was finally repealed in 1947. Restricting Asian migration by the dominant economies of the West was a theme right up until the 1960s.

"Scientific racism"—the use of pseudo-scientific techniques to suggest the inferiority of particular groups of people—was used to justify a policy of white expansion into the colonies during the Enlightenment period and, later, to introduce laws to exclude and discriminate against particular migrant groups. In part, as a result of the economic stagnation caused by two World Wars and the resulting exclusionary policies first introduced during the early twentieth century, there was an overall lull in global labor migration from about 1918 until 1945 (Castles and Miller, 2009).

During the Depression of the 1930s, some half million Mexicans were deported or pressured to leave the United States. When the demand for labor surged during the war years, Mexicans were recruited as temporary workers. Labor shortages occurred in many countries in the 1940s and 1950s, initiating the development of policies aimed at recruiting temporary workers. For instance, in 1942 the United States established a program to recruit Mexican laborers to work temporarily in agriculture, mostly in the southwest of the country. This program, known as the Bracero program, was designed to import Mexican workers during the harvest but was structured so that these migrants would return home once the season was over.

The Bracero program lasted until 1964 and an estimated 4.5 million Mexicans participated. As Kitty Calavita (1996, 289) states, "by the time the Bracero program ended, a relationship of symbiosis between Mexican immigrants and U.S. employers had become well-entrenched, facilitated and nurtured by more than 50 years of U.S. policymaking." Scholars have argued that irregular migration from Mexico to the United States really gained momentum with the ending of the Bracero program. Calavita (1996, 289) suggests "the employment of Mexican labor went underground as the guest workers of one era became the illegal immigrants of the next." Attempts to control migration from Mexico by the U.S. authorities have a long history. For example, in May 1954, U.S. Attorney General Herbert Browne announced that over the following months the U.S. Border Patrol would implement Operation Wetback, an intensive law enforcement initiative aimed at preventing increasing illegal border crossings from Mexico into the United States. By the end of that year, Browne announced that the border control had led to the apprehension and deportation of more than 1 million, mostly Mexican nationals (Hernandez, 2006). As we will discuss later in this chapter, control of the Mexican border is still a major element of migration policy in the United States today.

As we showed in Chapter 2, in the 1950s and 1960s, the German government recruited temporary foreign workers from Italy, Spain, Greece, Turkey, Morocco, Portugal, Tunisia, and Yugoslavia. These workers contributed to the rapid rise of Germany as an industrial power, working in low-skilled production jobs and, in later years, in the manufacture of electrical goods, automobiles, and textiles. Germany's Federal Labor Office (BfA) selected foreign workers on behalf of employers, tested their occupational skills, conducted medical examinations, and screened out those migrants with a criminal background. Migrants were granted permits to work only for restricted periods and only for specific jobs, and entry of any of the worker's dependents was forbidden.

The first recruits to the "guest worker program" (*Gastarbeiterprogramm*) were from European nations. With construction of the Berlin Wall and the resulting loss of East German labor, the program was expanded to include countries such as Turkey and Morocco. Turks became the largest group of temporary workers. Despite the fact that the agreement between Germany and source countries ended in 1973, many Turks, Italians, Greeks, Yugoslavians, and others remained in Germany. As with much temporary migration control, it was practically impossible to prevent family reunion and permanent settlement. Families of migrant workers established themselves, having children and building their lives in Germany. Starting out as temporary workers, these migrants and their families settled permanently in Germany.

It took until 2000, when Germany changed its citizenship law, for the country to finally recognize itself as a nation of permanent immigration. With the change in the law, citizenship could be granted to those children of migrants born in Germany, providing at least one of their parents has a permanent resident permit or has been residing in Germany for the past eight years. This legal shift from *ius sanguninis* (citizenship by descent) to *ius soli* (citizenship by birth in the territory) was an important milestone in German migration policy. However, one aspect of German citizenship law still impacts certain descendants of the *Gastarbeiterprogramm*. In the case of descendants from Turkey, for instance, they must choose their nationality by the time they reach the age of 23. Unlike migrants from the European Union, who can hold dual citizenship in Germany, Turks must choose to become German citizens and renounce Turkish citizenship or risk losing their German passports.

As we explored in Chapter 2, contract labor has always been an important element of migration. In recent times, the United Arab Emirates (UAE) has promoted the immigration of large numbers of workers from other countries through a system of work contracts. The Ministry of Labor issues a number of permits to labor-recruitment agencies to recruit foreign workers. These contracts severely restrict workers from bringing dependents such as wives or children to the UAE. Often, employers will keep the passports of foreign workers to prevent them from potentially switching jobs. Many of the recruited foreign workers in UAE live in worker camps, frequently in overcrowded conditions. These workers are often exploited, and employers do not hold adequately to their side of the labor contracts. In more recent years, there has been a change in migration policy in the UAE away from unskilled to more skilled migrants. In 1999, the nation introduced a ban on new permits and contracts for any people from India, mostly affecting unskilled workers. As we describe later, there is an increased emphasis among developed nations to regulate unskilled migration and encourage skilled migration.

In the postwar period, Australia launched an immigration policy, concerned that the population size of the nation made it vulnerable to invasion. Australia had a population of only 7.5 million at the time, in a country the size of the United States. The country had also just emerged from a war with neighboring Japan. The popular belief at the time was to "populate or perish." The focus of immigration policy in the early years after World War II was to encourage immigration from Britain—a continuation of the White Australia Policy. A large number of British did migrate to Australia through the Assisted Migration Scheme, known as the *Ten Pound Pom* program.

In the 1960s and 1970s, a number of legislative changes affected the nature of migration to developed countries. By the 1970s, the demographics in Australia began to shift in light of increasing migration and refugees from Asia and elsewhere. In 1972, the White Australia Policy was effectively abandoned and the country adopted a policy of multiculturalism. Trade relations between Australia and Asia expanded beginning around the 1980s. Asia's growth as an economic power has had a profound effect on the Australian economy. Financial, political, and cultural links between Australia and Asia have grown dramatically in recent decades. In the 1990s, there was some backlash to the migration in Australia, specifically in the area of asylum. Australia, like other developed nations, struggles with integration and acceptance of a diverse population of migrants.

As we mention in Chapter 2, in 1965 the United States introduced the Immigration and Nationality Act to end national origins criteria that had been part of immigration policy since the 1920s. Between the 1920s and 1965, immigration was restricted to a percentage of the mostly European foreign-born already in the country. With passage of the 1965 law, the preference system for migration to the U.S. became instead about family reunion and immigrant skills. Emerging during the civil rights movement, the new legislative policy was introduced as a way to correct inequities in the previous law. As we described in Chapter 2, the 1965 Immigration and Nationality Act had a tremendous impact on the demographics of the United States, allowing a surge of migration from Asia and Latin America.

More recent migration policy in the United States has turned toward recruitment of high-skilled migrants. An illustrative example is the U.S. Immigrant Act of 1990. The law was promoted as a compromise between exclusionary and inclusionary forces. There was concern about irregular migration throughout the Reagan years. Yet, the 1990 legislation was very liberal, increasing total admissions to the United States by some 40 percent, mostly of high-skilled workers. The United States turned its emphasis toward migration of high-skilled migrants. Australia and Canada had long emphasized migrant skills. In both of these countries, immigration legislation emphasizes the use of "the points system" where immigrants are selected based on characteristics such as age, education, language, and occupation. In Canada, the Immigration Act of 1976 was constructed such that, in the case of independent applicants for immigration status, assessment was based on points awarded for particular employment skills, education, and language abilities rather national origin. The United States' migration policy focused initially on national origins, then on family reunion and migrant skills.

Economies of the developed world have become increasingly dependent on migrant labor, both skilled and unskilled. In fact, migrant workers in developed nations tend to be concentrated in low-skilled jobs—those that domestic workers are unwilling to do. Examples are the concentration of Mexican migrant workers in the agricultural sector in the United States, or the use of Indonesian female migrants as domestic workers in Dubai. Structural changes in the global economy have led to growth in poorly paid service employment in developed nations, jobs that are typically filled by migrant workers. Developing migration policy suited to the recruitment of low-skilled workers has become a major challenge. Many low-skilled migrant workers are forced to work in the informal economy and are branded as illegal. In the 1980s and 1990s, developed nations increased their efforts to control migration as a political reaction to irregular migration. In the following section, we examine policy strategies specifically directed toward migration that aim to prevent irregular migration and monitor low-skilled workers.

IRREGULAR MIGRATION

The United Nations Population Division (United Nations, 1997, 27) states that irregular migration is "one of the fastest growing forms of migration in the world today." Almost every country across the globe reports the reality of irregular migration across their borders, including nations as diverse as Russia, the United Kingdom, the United States, South Africa, Botswana, South Korea, and Chile. Estimates suggest that there may be as many as 40 million irregular migrants worldwide, or about one-fifth of all global migrants (Düvell, 2011).

There are about 12 million irregular migrants in the United States (Warren and Warren, 2013), and some suggest that about 60 percent of these irregular migrants are Mexican (Passel and Cohn, 2010). There are estimated to be between 2 million and 4 million irregular migrants in the European Union (Vogel, 2009), although according to estimates by the Organization for Economic Cooperation and Development (OECD), this number was closer to 5 million in 2000. According to Koser (2004), more than one-half of all migrants in Africa and Latin America are thought to be irregular. Some estimates suggest that there are over 6 million irregular migrants in Russia (Düvell, 2011). The International Centre on Migration Policy believe that there are an estimated 2.5 million to 4 million migrants crossing international borders without the proper documentation each year.

Political concern about irregular migration has risen in recent decades, and there is a perception that irregular migration is out of control (Castles, 2004). Some argue that the rise in irregular migration indicates a "migration crisis" (Weiner, 1995; see Zolberg, 2001, for a critique of the term "crisis") but, as Khalid Koser (2007) suggests, the overwhelming majority of migrants are "legal" rather than "illegal." Most observers agree that most migrants are not irregular migrants. As Koser (2007, 59) points out, "the political significance of irregular migration far outweighs its numerical significance." As a political response, developed nations have formed a number of strategies to deal with what is perceived as a major problem.

LEGALIZATION

Policies to legalize irregular migrants have been introduced at different times by different countries. In the United States, the 1986 Immigration Reform and Control Act gave eligible irregular migrants the opportunity to become legal residents. The legislation stipulated a program to allow legalization of all migrants who could prove residence prior to January 1, 1982. Almost 1.7 million migrants applied for legal status, and about 97 percent of these applications were approved (Castles and Miller, 2009). In addition, there was a push to recognize family members of those seeking legalization. The United States introduced a "family fairness doctrine" to allow the Immigration and Naturalization Services to grant temporary legal status to undocumented dependents of legalizing migrants. Between 1986 and 2009, 3,760,618 migrants received legal status through various legalization programs, such as the Immigration Reform and Control Act (IRCA), in the United States (Kerwin, Brick, and Kilberg, 2012).

In the most recent debate on immigration in the United States once again there are serious discussions about legalization. As of May 2013, a bipartisan group of eight U.S. senators prepared to put forward an immigration bill providing a path to citizenship for 11 million undocumented migrants currently in the United States. The details of the plan are still under development, and it is difficult to determine if the U.S. Congress can successfully pass new immigration legislation during the Obama Presidency. However, it is likely that if legislation passes, it will include a form of a legalization program. Questions over deadlines for applications and eligibility criteria have been raised. There are politicians, traditionally opposed to legalization, being swayed by the strength of the Latino voting bloc who largely favor providing a path to citizenship for undocumented immigrants.

These political debates are global in nature. On February 8, 1980, following the evening news, a documentary on illegal Turkish garment-industry workers in Paris was aired. As Mark Miller (2002, 18) describes, "the program was shocking; tens of thousands of illegally employed aliens were working in abysmal conditions for wages that often were one third of the minimum wage." There was public outcry in France. Two days later, Turkish undocumented immigrants went on hunger strike, demanding legalization. Other hunger strikes and protests by foreign workers followed, and migrant mobilization helped lead to transformations in French migration policy.

Legalizations occurred in France at different times during the late 1960s and 1970s (Miller, 2002). In 1972, in a directive known as the *circulaire Fontanet*, the French government restricted access to legalization to only the most skilled irregular migrants. In 1981, the French Socialist government declared changes to the legalization program, making it easier for a broader group of migrants to gain legal status. With the support of trade unions and immigrant organizations, immigrants mobilized through hunger strikes and protest and, in the end, some 120,000 to 150,000 applicants gained legal status with passage of the 1981 law (Castles and Miller, 2009).

French migration policy tightened when Nicolas Sarkozy became Minster of the Interior in the mid-2000s. Deportation of undocumented migrants was increased, both during his time as Minister of the Interior and later during his presidency. In 2006 protests ensued when Sarkozy sought to deport the children of undocumented migrants at the end of the French school year. That same year, Sarkozy developed national immigration legislation, adopted on July 25, 2006, that aimed to encourage high-skilled migration, restrict migrant family reunion, and introduce procedures to limit applications for undocumented migrants to gain legal status. The more recent French Socialist government has struggled with the issue of migration and has not yet offered to loosen migration control. More broadly, current economic conditions in France and Europe have influenced the immigration debate. In light of difficult economic times, the Minister for the Interior, Manuel Valls, announced reductions in financial benefits to French immigrants in early 2013.

SUBURBAN IMMIGRANTS OF PARIS, FRANCE

Clichy-sous-Bois is an inner-ring suburb located about 12 miles east of downtown Paris. It is a town of approximately 80,000 residents. The suburb has historically been composed of poor and working-class neighborhoods. In particular, in recent decades, non-European migrants have increasingly settled in Clichy-sous-Bois. By the beginning of the twenty-first century, nearly one-third of residents were foreign-born—at least 25,000 in Clichy-sous-Bois. A large number of these migrants arrived from former French colonies in northern Africa, and these migrants are largely Muslim. In total, 5 million Muslims live in France today. Many of Paris' other eastern inner-ring suburbs mirror the socio-economic structure of Clichy-sous-Bois, including Tremblay, Sevran, Blanc-Mesnil, Clichois, and Neuilly-sur-Marne. In many of these communities, residents suffer from high levels of unemployment, unsafe neighborhoods, poor schools, and little or no public support of infrastructure or social services.

Tensions in these communities mounted for some time until October 27, 2005 when police arrested a group of teenagers for attempted robbery. Reports of brutality were asserted after one of the teenagers died. This and other events triggered many riots in surrounding communities (Crampton, 2005). Residents reported to the press that the riots were the result of poor economic conditions, combined with police harassment and poor public services. "People are joining together to say we've had enough. We live in ghettos. Everyone lives in fear," said one protester. The social and physical damage was significant. Two were killed, some 9,000 cars were burned, 3,000 residents were arrested, and buildings in more than 250 towns were damaged or destroyed.

The civil unrest prompted a strong political response. Politicians and political groups quickly mobilized. Jean-Marc Ayrault, the president of France's Socialist Party, said, "we see that the situation in certain neighborhoods is not getting better at all but degenerating." He blamed the conservative government for the problems. In contrast, others strongly disagreed. Philippe de Villiers, a politician in the French Parliament, said that his country needed to, "stop the Islamization of France." He blamed the riots on the "failure of a policy of massive and uncontrolled immigration" (Keaten, 2005). Meanwhile, the French government adopted a "zero-tolerance" policy for urban violence and riots in these communities. Ultimately, the French Parliament declared a state of emergency on November 8, 2005, which lasted for three months.

The case demonstrates the perils of not fully incorporating migrants into society—socially, economically, and politically. While there are multiple policy approaches to achieving incorporation, we see that in the case of Paris, the failure to incorporate disenfranchised migrants had serious consequences. The debate continues today.

In 2008, the European Pact on Immigration and Asylum originally stipulated that European Union members would only allow legalization of undocumented migrants on a case-by-case basis, rather than through the use of broad-based legalization programs. Irregular migrants legalized in one European Union member state are, under EU rules, allowed to travel and in some cases reside in other member states. As Donald Kerwin, Kate Brick, and Rebecca Kilberg (2012) suggest, the majority of recent legalization programs are found in southern European countries such as Italy, Spain, Greece, and Portugal. These researchers show that the overwhelming majority of applications to EU legalization programs over the ten years from 1997 to 2007 occurred in Italy, Spain, and Greece. These countries implemented legalization programs in light of increased undocumented migration from north and sub-Saharan Africa. In contrast, northern European countries have resisted the implementation of broad-based legalization programs and have encouraged southern European countries to do the same in light of European rules on integration.

Those who object strongly to legalization programs argue that they encourage further illegal migration and undermine the rule of law. Those who favor legalization programs suggest there are social and economic benefits. They posit that moving irregular migrants from the informal to the formal economy helps increase wages and eliminates unfair advantages to employers who hire undocumented migrants. They also suggest that legalization increases tax payments, thus benefiting society at large. In addition, legalization is thought to benefit immigrant communities: when some members of an immigrant family or community are legal and others are not, the family and community are divided. Undocumented immigrants take a risk when returning home to visit family and friends. Legalization helps with immigrant integration, family unity, and career development.

EMPLOYER SANCTIONS

Since the 1980s and 1990s, legislative and policy responses to irregular migration, particularly in Europe and the United States, have included approaches to punish employers for hiring undocumented or unauthorized migrants (Castles and Miller, 2009). These employer sanctions have been met with some resistance from industry, and their effective enforcement has proven difficult. For example, in the case of the United States, employer sanctions did not exist until after the passage of the 1986 Immigration Reform and Control Act (IRCA). This law made it a punishable offense for employers to hire undocumented migrants, requiring them to complete an "I-9 Form" for the hiring of new employees. Typically, the punishment for employers is a fine.

The immigration policy regime in the United States was restructured shortly after the attacks on September 11, 2001. Immigration services were placed under the auspices of the Department of Homeland Security. This department, through the U.S. Citizenship and Immigration Services, manages an Internet-based system called "E-Verify" that allows employers to verify the migration status of potential and hired employees. This system, created in 1997 and grown considerably since, has become a tool in monitoring potential illegal hiring practices. All federal agencies and federal contractors must use it. Employers enter basic information about their employees into the online system and cross-check to verify that the employees are legally authorized to work. If an employee is found to be unauthorized, the employer is required to fire them.

MIGRATION AND THE QUESTION OF SCALE

Over the years, there has been bilateral agreement between countries over migration. Examples mentioned above include Germany's guest worker program and the Bracero program between the United States and Mexico. Beyond bilateral agreements, there has also been regional integration, allowing for the freer movements of people across borders within a region. A case in point is the European Union. Transnational regional entities have evolved, with implications for migration. At the same time, there is evidence that migration policy in the United States is devolving to lower levels of government as a result of frustration at the inability of the federal government to agree on a unified and politically satisfying strategy toward immigration. In the following sections, we will examine, on the one hand, regional integration and free movement of people in the European Union and, on the other hand, devolving governance of immigration in the United States. Migrant regulation can take place at different scales.

REGIONAL INTEGRATION IN EUROPE

In 1951, the Treaty of Paris was signed, creating the European Coal and Steel Community (ECSC). The Treaty prevented restrictions on employment in these industries among the six member nations of Belgium, France, Italy, Luxembourg, the Netherlands, and West Germany. Later, the 1957 Treaty of Rome established the European Economic Community (EEC), a common market between those same six states. Under Article 48 of the Treaty of Rome, workers from member states could move freely to another member state for employment. In 1992, the Maastricht Treaty established the European Union, which was signed by the original six member states along with the additional states of Denmark, Ireland, the United Kingdom, Greece, Portugal, and Spain. Since the Maastricht Treaty, the European Union has expanded to now include some 26 European states. Under the European Union, member states are part of a single market where there is free movement of people, goods, services, and capital.

Over the timeframe of European integration, there is one agreement that most directly impacts regional migration. This is the Schengen Agreement, signed in 1985 by France, Germany, Belgium, Luxemburg, and the Netherlands. The agreement sought to establish a border-free Europe in which citizens of the European Union could move freely within the region while border controls for non-European Union citizens were maintained. There was controversy over this agreement, with many nations and groups opposed to the loss of state control over migration across their borders. Finally, in March 1995, the Schengen Agreement came into force for those states that were signatories: Germany, Belgium, Spain, France, Portugal, Luxemburg, and the Netherlands. At first, the United Kingdom and Ireland refused to sign, insisting on their own border controls, but later these nations also agreed to take part in some aspects of the Schengen Agreement.

At various times, as the European Union has expanded, there has been disagreement around labor mobility and migration. For instance, between 2000 and 2001 when countries of central and eastern Europe were seeking membership of the EU, there was concern, especially in France and Germany, that workers from Poland in particular would flood their labor markets. Eastern and central European countries finally joined the EU in 2005. Initially, many of the 15 member states placed labor restrictions on workers. Eventually restrictions were lifted, yet national governments consistently demonstrated concern for the impact of labor migration on domestic workers and employment. At various points in time, the Schengen Agreement has been a source of political conflict in the public discourse on migration. For example, to help his presidential campaign, Nicolas Sarkozy threatened to withdraw France from the Schengen Agreement if significant measures were not put in place to deal with irregular migration in Europe. Open borders for European Union workers makes it easier for workers beyond the EU to enter without the proper documentation. As Europe deals with its latest economic crisis, questions about regional integration and its effects are likely to continue.

DESCALING OF MIGRATION POLICY IN THE U.S.

In recent years, there has been a proliferation of subnational regulation designed to control the movement of immigrants in the United States. For instance, in April 2010, Governor Jan Brewer of Arizona signed into law one of the toughest policies on immigration control, which allowed police to arrest immigrants who cannot prove their lawful entry into the country. One of the primary goals of the legislation, known as SB 1070, is to identify, prosecute, and deport undocumented immigrants. By July 2010, the U.S. Department of Justice had filed a lawsuit challenging the constitutionality of Arizona's immigration law. The federal lawsuit asked the U.S. District Court in Phoenix to uphold the federal government's "plenary power" over the formulation and enforcement of immigration, and rule that Arizona's new law violates the supremacy clause of the U.S. Constitution. Upon filing the lawsuit, Attorney General Eric Holder stated, "Setting immigration policy and enforcing immigration laws is a national responsibility. Seeking to address the issue through a patchwork of state laws will only create more problems than it solves." Ultimately, the U.S. Supreme Court was asked to review the constitutionality of Arizona's legislation. In June 2012, the Court delivered a split decision, upholding some and blocking other parts of the law. The Court dismissed the element of the law that made it a crime for undocumented immigrants to be present in and seek employment in Arizona. The most debated provision of the law, that required state law enforcement officers to determine the immigration status of people that they arrest or suspect of being an undocumented immigrant, was upheld.

Arizona is not alone in its tough stance on immigration. Other states have contemplated and even developed their own versions of this type of legislation. Certainly, the Supreme Court decision on SB 1070 is likely to have implications far beyond Arizona. According to a study by the National Conference of State Legislators, the five states of Alabama, Georgia, Indiana, South Carolina, and Utah all enacted legislation similar to Arizona's in 2011. But the federal government has filed complaints on immigrant enforcement legislation in three of these states—Alabama, South Carolina, and Utah—and there are still pending class-action lawsuits regarding the constitutionality of Arizona's SB 1070. The trend may be of an increasing involvement of states in immigration reform, but the debate over who has the power to enact this reform is still ongoing.

States are not the only subnational battleground for immigration control. Small towns, suburbs, and cities across the United States have become active in the immigration debate, developing a host of both anti- and pro-immigration-related legislation. In 2006, the Fair Immigration Reform Movement compiled a database of local government immigration-related legislation and found a total of 135 "anti-immigration" measures that have been introduced at the local jurisdictional level. These include passing ordinances to fine landlords who rent to undocumented immigrants, enforcing housing code violations targeting overcrowding by immigrants, preventing establishment of day laboring sites, and, of course, ensuring immigration control by requiring local police officers to check the immigration status of detainees or, in some cases, of local residents stopped for traffic violations.

Probably the most renowned local exclusionary immigration legislation in the United States is that introduced by the community of Hazleton, Pennsylvania in July 2006. At that time, the city council in Hazleton passed the Illegal Immigration Relief Act (IIRA) by a vote of four to one. The ordinance had a number of elements aimed specifically at businesses, landlords, and social services that potentially worked with undocumented immigrants. The ordinance sought to deny licenses to commercial entities that employed undocumented immigrants, to fine landlords US$1,000 for each undocumented immigrant housed in his or her rental property, and to declare English the official language of Hazleton. The mayor, Lou Barletta, wore a bullet-proof vest on the day of the signing, stating at the time "[to] the illegal citizens, I would recommend they leave … what you see here tonight, really, is a city that wants to take back what American has given it" (Associated Press, 2006).

In 2007, a federal judge struck down Hazleton's IIRA as an unconstitutional ordinance. The law has not been enforced since it passed the city council, and it remains in legal limbo, in part because elements of Arizona's SB 1070 were struck down. Hazleton's IIRA inspired similar ordinances to spring up across the country. As Lou Barletta stated back in 2006 (Rubinkam, 2011),

> Hazleton has paved the way for other cities and states across the country to enact similar laws, so this is a great day for all of those cities and states, and for the people of Hazleton who had to endure criticism from those who opposed what we were trying to do because the federal government didn't want to do its job.

The weighty elements of the narrative prior to passage of Hazleton's IIRA focused heavily on the notion of security. During the debate, local politicians and certain constituents in Hazelton were focused on the undocumented immigrant as a perceived threat, a physical threat (hence Lou Barletta's bullet-proof vest on the day of signing), and an economic threat to a certain way of life.

Much of the legal research examining the evolution of exclusionary policies at the state and local level concludes that the federal government should continue to have primary authority over the regulation of immigration (Harnett, 2007). Legal scholars and activists have been and continue to be concerned that exclusionary policies violate the civil rights of immigrants and others. Many local ordinances have been found to violate "due process" rights of landlords and renters, for instance (Esbenshade, 2007). On the other hand, some legal scholars suggest that local immigration policy-making may be a necessary step since often local governments are required to enforce immigration law. Such requirements are viewed as unfunded mandates that undermine local control and self-government (Parlow, 2007). Recent legal wrangling over local immigration regulation highlights the scalar tension in the policy arena of immigration reform. Some in the legal community advocate that federal exclusivity in the arena of immigration should be reformulated. This alternative view offers a functional account of subfederal immigration regulation by suggesting that the federal–state–local dynamic should operate as an integrated system to manage contemporary immigration (Rodriguez, 2007).

Noting the "down-shifting" of immigration reform, scholars have argued that this devolution began in earnest with "the new federalism" model first introduced by Reagan in the 1980s, and then further amplified by new laws introduced in the mid-1990s, which limited non-citizens' access to welfare and medial programs (Varsanyi, 2008). These laws and strategies also included an expansion of the array of criminal offenses for which non-citizens could be deported. Local service providers and, in some cases, private citizens were encouraged to engage in enforcement of immigration law by neoliberal policies. For instance, the Immigration Reform and Control Act of 1986 introduced an employer sanctions program aimed at penalizing employers for hiring immigrants unauthorized to work, and, as a result, employers became responsible for checking and maintaining immigrant records of their employees (Ridgley, 2008). This same act led to an increase in immigration raids in various cities, conducted at times with the help of municipal police and local officials. The Illegal Immigration Reform and Immigrant Responsibility Act (IIRIRA) of 1996 allowed local police officers to receive training in immigration control from Immigration and Customs Enforcement (ICE) (Coleman, 2007).

Welfare reform, the IIRIRA, and other federal-level policies passed the responsibility of immigrant service provision and enforcement of immigration control to states and local communities. Subnational governments engaged in immigration reform as a response to neoliberal federal policy. In the minds of certain local and state governments, the federal government did, and continues to do, little to discourage the influx of large numbers of undocumented immigrants and, at the same time, has minimized service provision for immigrants at the local community level (Ellis, 2006). The removal of state-centered welfare of new arrivals to the United States is an effect of the neoliberalist framework that depletes the state's function as universal service provider. Under neoliberalism, the state's main function is to ensure optimal functioning of the market and secure private property rights (Harvey, 2007). Yet, the prominence of market efficiency over distribution and service provision does not mean minimization of government intervention. Rather, it suggests that the role of the state has been reconstituted to also include new forms of social control (Peck and Tickell, 2002). This is clearly demonstrated by recent engagement in the policing and regulation of immigrant communities.

DISPLACEMENT AND MIGRATION POLICY

Attempting to control and regulate population displacement became paramount with the outbreak of World War I (Moses, 2006). The war caused great disruption, and many people were forced to migrate out of fear of persecution. States were reluctant to accept forced migrants. For instance, when the war broke out, more than 5 million migrants from Russia and Austria-Hungary traveled through Germany on their journey to ports bound for the United States. These migrants were carefully monitored. As transit migrants crossing German territory, they traveled in special trains and special train compartments. These trains were strictly controlled to prevent emigrants from Russia and Austria-Hungary remaining in Germany. The strict "transit migrant control," known as *Durchwandererkontrolle*, was aimed to prevent any potential irregular immigration to German territory (Bade, 1995). How to manage refugee populations was of major importance during this period.

After the Bolshevik Revolution in 1917, some 1 million refugees fled the newly formed Communist regime of Russia. The Soviet decree of 1922 denationalized many of these refugees, leaving them stateless (Moses, 2006). In 1921 Fridtjof Nansen of Norway was appointed by the League of Nations to be High Commissioner for Russian Refugees. In this role, Nansen defined the legal status of stateless refugees, primarily from Russia, by developing the Nansen passport system. During World War I, European states introduced border passports to keep out spies and to control the out-migration of skilled labor. States within Europe were unwilling to return to the pre-war times of relaxed border security. Introduced in 1922, Nansen's passport for undocumented refugees was recognized by many nations at the time. Unfortunately, though, these documents did not necessarily persuade states to, as Moses suggests (2006, 52), "[take] any responsibility for the passport holders. Stateless refugees now had stateless passports, but their plight remained dismal."

The extent of the refugee crisis became worse after World War II. There was a stronger international commitment to dealing with the crisis that emerged after Nazi expansion and the reconfiguration of Europe in the aftermath of the war. Some estimates suggest that there were 30 million displaced people in Europe after World War II, and that 11 million of them were outside their country of origin (Moses, 2006). The United Nations emerged as the leading international organization to deal with this problem. In 1943, the United Nation's Relief and Rehabilitation Administration (UNRRA) was established, followed in 1946 by the International Refugee Organization (IRO). Both were succeeded by the creation of the United Nation's High Commissioner for Refugees (UNHCR) in 1950. UNHCR still exists today as the primary international organization to address the concerns of refugees.

Persecution during World War II, and the brutal reality of the Holocaust in particular, resulted in international concern for basic human rights. In 1948, the Universal Declaration of Human Rights was adopted. Article 13 of this declaration recognizes everyone's right to leave any country, including his or her own. In 1985, the UN General Assembly adopted the Declaration on the Human Rights of Individuals Who Are Not Nationals of the Country in Which They Live. This declaration explicitly acknowledges that

> nothing in this declaration shall be interpreted as legitimizing any alien's illegal entry into and presence in a State, nor shall any provision be interpreted as restricting the right of any State to promulgate laws and regulations concerning the entry of aliens and the terms and conditions of their stay.

It should be noted that despite willingness to ensure the right of an individual to leave his or her country of origin, there is no corresponding mandate that nations must allow them entry. The nation-state has ultimate authority over who is and is not allowed enter that state. As Gil Loescher and James Milner (2011, 190) suggest, "international contemporary concern for refugees, centered around the concepts of international protection and human rights, has its origins in the immediate aftermath of the Second World War." The World War II conflict heavily influenced the way refugee policy evolved, bringing together international concern about human rights and the emerging reality of forced migration.

In 1951, delegates from 26 countries adopted the United Nations Convention relating to the Status of Refugees, a key legal document defining who a refugee is, their rights, and the obligations of states that sign the convention. In 1965, at a colloquium in Bellagio, Italy, some 13 legal scholars gathered to consider ways to improve upon the 1951 Convention. They drafted the 1967 Protocol relating to the Status of Refugees, removing the geographical and temporal restrictions laid out in the 1951 Convention. A refugee is defined by the 1951 United Nations Convention as a person who resides outside his or her country of nationality and who is unable to return because of "well-founded fear of persecution on account of race, religion, nationality, membership in a particular social group, or political opinion." This sentence recalls the concern for non-discrimination in the Universal Declaration of Human Rights of 1948.

More recent criticism of the 1951 United Nations Convention is that it does not include sex or sexual orientation among the various characteristics on which people can be persecuted. In nation-states such as Iran, Mauritania, Nigeria, Saudi Arabia, Sudan, and Yemen homosexuality is illegal and can be punishable by death. Moreover, the persecution of gay, lesbian, and transgendered people occurs in many countries, particularly more conservative ones in the Middle East and Africa. In a similar manner, women can experience persecution on the basis of their gender. We explored in more detail issues related to migration and gender and migration and sexuality in Chapter 3.

According to UNHCR, by the end of 2011, worldwide there were 42.5 million people forcibly displaced in large part because of conflict or persecution. Table 5.1 lists the ten main source countries of forcibly displaced populations. Refugees are forced migrants, but

Table 5.1 Ten main source countries of forcibly displaced people

Source country	*Forcibly displaced population*
Afghanistan	2,664,400
Iraq	1,428,300
Somalia	1,077,000
Sudan	500,000
Democratic Republic of the Congo	491,500
Myanmar	414,600
Colombia	395,900
Vietnam	337,800
Eritrea	252,000
China	205,400

Source: The United Nations High Commissioner for Refugees (UNHCR) Global Trends 2011.

it is important to recognize that not all forced migrants are refugees. Many people are forced to migrate but for reasons not recognized by international refugee law. Since many refugees remain in their country of origin, they are not classified as international refugees. A refugee, by definition, must cross international borders. As of the end of 2011, there were 10.4 million refugees globally.

There are currently 145 states that are party to the 1951 United Nations Convention and 146 states signed the 1967 Protocol. States that are party to the Convention accept the principle of *non-refoulement*, recognizing that no refugee should be returned to any country where he or she would be at risk of persecution. Article 3 of the 1984 Convention Against Torture extends the same protection for those who believe they may be subject to torture upon return. The non-refoulement principle has become important in consideration of refugee resettlement and determining the status of those individuals seeking asylum. Asylum seekers, as people waiting for a decision on refugee status, might be endangered upon return to their country of origin.

The global refugee population has risen dramatically in the past 40 years. In 1975, there were 2.4 million refugees; by 1990, this population had grown to 14.9 million. In 1993, after the end of the Cold War, there were 18.2 million refugees, increasing to 18.5 million in 1995. Since then, there has been some decline in the number of

Table 5.2 Top ten countries where refugees reside

Country	*Refugees*
Pakistan	1,702,700
Islamic Republic of Iran	886,500
Syrian Arab Republic	755,400
Germany	571,700
Kenya	566,500
Jordan	451,000
Chad	366,500
China	301,000
Ethiopia	288,800
United States	264,800

Source: United Nations High Commissioner for Refugees (UNHCR), Global Trends 2011.

refugee migrants. According to UNHCR, there were an estimated 13.8 million refugees in 2005, increasing to 16.3 million by 2010. Table 5.2 lists the top ten countries where refugees reside.

Beginning in the 1960s and 1970s, rebellions and conflicts following decolonization and independence for many African countries generated vast numbers of refugees across the continent. With decolonization, borders were often drawn in ways that did not respect ethnic or tribal differences. Conflicts resulted, and many fled for fear of persecution. Descendants of European settlers also became refugees, especially in places such as the Congo, Mozambique, and Angola. For instance, nearly 1 million Portuguese, or settlers of Portuguese descent, fled Mozambique and Angola as refugees in the mid-1970s.

The large increase in the number of refugees during the 1990s was the result of a number of major conflicts, and, in addition, dramatic changes in Europe with the breakup of the former Soviet Union. Some prominent conflicts during this period include the Bosnian War between 1992 and 1995, and the conflict between the Hutu and Tutsi groups in Rwanda in 1994. A feature of both wars was the policy of ethnic cleansing, a process by which one ethnic or religious group aims to eliminate from their territory another unwanted ethnic or religious group, either by forcible displacement or by mass murder. In the case of the Bosnian War, Croat and Serb forces developed policies to rid their territories of groups other than themselves. Mass rape of women also occurred during the Bosnian conflict. In the case of Rwanda, there was a large massacre of Tutsi by the Hutu during the Rwandan genocide. The Tutsi then took control of the country during a civil war and, fearing reprisal, about 2 million Hutu fled to neighboring African countries.

In more recent times, forced migrations have also occurred. For example, millions fled the Darfur region of Sudan in 2007. Similarly, there was an exodus of people from Iraq following the U.S.-led invasion in 2003. Also, beginning with the Soviet invasion in 1979, and the later invasion led by the U.S. in 2001, millions of Afghans fled their country, some returning but many still residing in neighboring Pakistan and other nations. Other countries that have experienced large-scale refugee crises include nations such as Mozambique and Namibia, Myanmar, and most recently Syria. As Khalid Koser (2007) suggests, since the time when the global community became involved in monitoring and helping refugees shortly after World War II, the geography of refugee flows has changed dramatically. Many refugees originate from some of the poorest parts of the world, most especially Africa. Significant numbers of refugees end up in developed parts of the world, particularly the United States. But often, the responsibility to host these refugees falls on those nations neighboring refugee-producing countries (see Table 5.2). This is what Hathaway and Neve (1997, 141) refer to as an "accident of geography." Today, there are an estimated 2.3 million Syrian refugees who fled the crisis in their country, many to the nearby countries of Lebanon, Jordan, and Turkey.

EVOLUTION OF UNHCR AND THE INTERNATIONAL REFUGEE REGIME

As the geography of refugee flows has changed, so too has the work of the UNHCR. In his book, *The UNHCR and World Politics*, Gil Loescher (2001) examines the evolution of the UNHCR and its role in managing refugees across the world. The UNHCR was initially established as a temporary organization, given little funding, and expected to finish its work in three years. Today, the UNHCR exists as an agency with a budget in 2010 of US$3 billion. Launched with a budget of US$300,000 in 1950, it has grown extensively. It is funded almost entirely by voluntary donations, with 93 percent of these donations coming from governments. According to the UNHCR, the top five donors in 2008 were the United States, the European Commission, Japan, Sweden, and the Netherlands.

There are other organizations that assist the UNHCR as part of an international refugee regime. This term is used to denote the institutions internationally that aid refugees, along with the various human rights laws that protect them. These institutions include, for instance, the International Organization for Migration (IOM). The IOM provides data and statistics related to migration. But this organization also offers logistical aid in the case of refugees, especially in the area of transportation. The IOM offers resettlement assistance, working closely with the UNHCR and governments, offering case processing, health assessments, and transportation services to help in resettling refugees. There is also a wide range of non-governmental organizations that assist the UNHCR in its refugee work, organizations such as Catholic Relief Services, the International Committee of the Red Cross, the World Food Program, the United Nations' Children's Fund, Lutheran Immigration and Refugee Services, the International Rescue Committee, OXFAM, among a host of others. These organizations often take responsibility for managing refugee camps, and distributing food, medical services, and education to refugees.

During the Cold War period, the West was willing to offer asylum for individuals interested in defecting from communist countries. This was an excellent propaganda tool for Western countries. At the height of the Cold War period, UNHCR focused its efforts on providing legal protection to refugees fleeing countries such as Russia, Hungary, Czechoslovakia, and East Germany. The UNHCR was involved in helping after an outpouring of refugees from Hungary following the Soviet invasion in 1956.

Beginning in 1960s and 1970s, the UNHCR became more involved in refugee situations in the global South, which at times has been a region of crisis and conflict (Loescher and Milner, 2011). Since the 1980s, the UNHCR has not only shifted its geographic focus, but it has also moved beyond being merely an agency providing legal protection to refugees. The UNHCR now views itself as an institution providing humanitarian assistance not only to refugees but, beginning in the 1990s, also to victims of natural disasters and Internally Displaced People (IDP) who have never crossed international borders. As Gil Loescher and James Milner (2011, 194) declare, the expansion of the UNHCR's role has been controversial, with some concern that the institution is being "used by states in ways that may contradict or undermine refugee protection."

The UNHCR is a major player in the regulation of refugees, offering assistance and guidance to states on refugee matters. States often defer to the UNHCR on legal questions and concerns about asylum and refugee status. As Gil Loescher and James Milner (2011, 194) state, "For most of its history, [the UNHCR] has acted as a 'teacher' of refugee norms." However, the agency has no direct control over what states do. It can advise, but ultimately, states remain "the predominant actors" in the control and supervision of displaced populations (Loescher and Milner, 2011, 194). This is especially the case in questions of asylum.

ASYLUM SEEKERS AND PROTRACTED REFUGEE SITUATIONS

During the 1980s and 1990s, large numbers of asylum seekers came directly to Europe and North America, seeking refuge from conflicts that occurred in various nations. For example, with the dissolution of the Soviet Union and the wars resulting from the dismantling of the former Yugoslavia, asylum applications soared in countries such as Italy, Germany, and the United Kingdom. Between 1991 and 1995, some 1.3 million asylum seekers arrived in Germany alone. At times, asylum seekers arrived in Europe and North America from distant countries such as Afghanistan, Somalia, and Sri Lanka.

Asylum became highly politicized in the 1990s. Many of the asylum seekers arrived at borders without any authorization, as "spontaneous asylum seekers." In this way, these migrants were very different to refugees. They came needing to prove their refugee status. European and North American countries responded with a series of restrictions. Changes were made to national legislation to restrict access to refugee status. "Carrier sanctions" were introduced, compelling airlines to check for the correct documentation before allowing people to embark. Policies were introduced by European nations to return asylum seekers to those nations used as transit routes—to so-called "safe third countries" (Castles and Miller, 2009). Asylum seekers were denied access to welfare benefits. One of the major political arguments used against admitting asylum seekers is that they are, in fact, labor migrants and not refugees. Proponents of this argument suggest that many asylum seekers are really irregular migrants. The term "migration-asylum nexus" emerged to describe the policy challenges involved in distinguishing refugees from phony asylum applicants, genuine asylum seekers, and irregular migrants (Koser, 2007). It can be a difficult balance, but it is ultimately nation-states that determine the status of asylum seekers once they arrive spontaneously at their borders.

Many developed nations are becoming less willing to admit asylum seekers. In the United Kingdom, for instance, only 10 to 20 percent of asylum seekers satisfied the criteria of the 1951 United Nations Convention and were granted refugee status in the late 1990s and early 2000s. Australia has recently been lambasted for its detention and treatment of asylum seekers arriving by boat to its shorelines. Many, coming most recently from Sri Lanka, are detained on small islands off the coast of Australia, awaiting determination of refugee status. As it becomes more difficult to be admitted to nations of the developed world, asylum seekers increasingly pursue refuge in nations such as South African, Kenya, Egypt, and Thailand (Castles and Miller, 2009) or, like the majority of refugees, flee to nearby nations.

MIGRATION IN IRELAND

Throughout history, the population of Ireland has ebbed and flowed as the global economy fluctuated. From the days of the Irish famine to the arrival of economic expansion in the twenty-first century, the Irish population has been distinctly susceptible to larger global migration patterns. From 1845 to 1849, the Great Famine of Ireland claimed the lives of 1 million residents and forever changed the landscape in this small nation. The impact was dramatic. The population declined from a peak of about 8 million in 1840 to a little over 4 million in 1920. It was not until the 1970s that the population began to grow again.

Later, during the global economic recession of the 1980s, many Irish again fled to the United Kingdom, Australia, Canada, and the United States. In 1989, at the peak of the recession, some 70,000 Irish residents, or 2 percent of the nation's population, migrated to other countries. This trend of population loss persisted until 1995, when net migration in Ireland almost balanced; that is, the number of emigrants equaled the number of immigrants. In the subsequent decade, approximately 276,000 people migrated to Ireland.

This new migration was not accidental. Ireland pursued a two-tiered approach to implement migration policies to grow its population. First, it welcomed back returning Irish nationals. People who had left Ireland were free to return home to settle and work. Second, the nation expedited visas and authorizations for foreign nationals to work in Ireland. This facilitated the bureaucracy of obtaining permits to work in Ireland and attracted a high-skilled labor force. The results have been dramatic. Over 100,000 new work permits and just as many renewal permits were issued in the first five years of the 2000s. The majority of the permits were issued to migrants from Poland, the Philippines, Lithuania, Ukraine, and Romania. The addition of new EU member nations and ease of movement within the EU facilitated this growth.

At the same time, Ireland also began to discourage refugees and asylum seekers. In 1992, there were 39 asylum applications to Ireland, but a decade later, there were more than 11,000 applications (Gilmartin, 2008). The majority of the applications were from Nigeria and Romania; others were from Moldova, Somalia, and Ukraine. Paul Cullen (2000) advances the idea that Ireland is an attractive place for asylum seekers for numerous reasons. He notes that Ireland, and the arrival of the economic success of the "Celtic Tiger," offers the promise of a better life. Ireland has a more generous social welfare system than the United Kingdom. There is also less racism in Ireland than in other EU member countries. Immigrants viewed these conditions as favorable, and Ireland thus emerged as an immigrant gateway at the beginning of the twenty-first century (Gilmartin, 2008). While the number of asylum seekers and refugee applications has declined, Ireland has faced many challenges of integrating and assimilating migrants into its society.

Ireland's migration history is nothing but turbulent and resilient. Dramatic population changes have closely mirrored the growth and decline of the global economy. Once a strictly homogenous people and country, today's Ireland reflects a patchwork of migrants from many parts of the world.

Many times refugees spend years in refugee camps with no prospect of resettlement or return. Refugee resettlement occurs when the UNHCR, in cooperation with governments of resettlement countries—mainly the U.S., Canada, Australia, and New Zealand, and to some extent members of the European Union—offers long-term protection and assistance. This is especially in cases where refugees are endangered if they are sent back to their country of origin. Refugee resettlement within developed nations has declined in recent years. The UNHCR recognizes "protracted refugee situations" where refugee populations of 25,000 or more remain in exile for a period of five years or more (Castles and Miller, 2009). At the end of 2009, there were an estimated 10.3 million people living in protracted refugee situations, which constituted two-thirds of the refugee population worldwide. The UNHCR identified 30 countries, primarily located in Africa and Asia, where protracted refugee situations exist.

HUMAN SMUGGLING AND HUMAN TRAFFICKING

A disturbing feature of contemporary migration is the rise of organized smuggling and trafficking of migrants for profit. First, it is important to distinguish between smuggling and trafficking. In the case of smuggling, migrants are moved illegally for profit and, although unequal partners in the transaction, the move is voluntary and the profit comes from the move itself. In the case of trafficking, as Ann Gallagher (2002), of the UN High Commission for Human Rights states, "the movement … is based on deception and coercion whose purpose is exploitation. The profit in trafficking comes not from the movement but from the sale of the trafficked person's sexual service or labour in the country of destination." Sometimes the line between human smuggling and human trafficking is blurred. For instance, in a situation where a migrant does not pay the smuggler until he or she arrives in the destination country, he or she will be in debt to the smuggler. As Khalid Koser (2007, 65) suggests, this "opens up the possibility of exploitation [once they arrive]." In the following sections, we describe the relationship between human smuggling and border control, and the policy environment of human trafficking.

POLICY ENVIRONMENT AND HUMAN TRAFFICKING

The U.S. State Department provides estimates on the number of people trafficked annually. According to a 2007 report, an estimated 800,000 people are trafficked across borders each year; if trafficking within nations is included, the numbers increase to between 2 and 4 million people annually. The International Labor Organization (ILO) estimates that in 2005 there were 2.45 million victims of internal as well as transnational trafficking (Belser, De Cock, and Mehran, 2005). According to estimates from 2004, two-thirds of victims were trafficked within Asia and Europe.

Trafficking takes different forms. Some victims, mostly women and girls, are forced into prostitution in large cities and sex tourism areas. Another place where these victims end up is around military bases. Many men, women, and children become slaves or forced into bonded labor. Mostly women and girls are forced to be domestic servants. Children in particular are forced to become beggars on the street or, worse, child soldiers. According to the U.S. Department of State (2007), 80 percent of victims of human trafficking are women and young girls.

Controlling human trafficking has proven difficult, and certainly more attention to the issue is needed to end this insidious and despicable exploitation of people. Measures are needed at the international scale to provide some protection to the victims of trafficking. Foremost, in 2000, the Palermo Protocols were adopted by the United Nations. The "Protocol to Prevent, Suppress and Punish Trafficking in Persons, especially Women and Children" provides definitions and sets guidelines for trafficking crimes. The United Nations Office on Drugs and Crime (UNODC) is responsible for implementing the Protocol. As part of the Protocol, nation-states must take specific actions to prevent trafficking and prosecute traffickers. The UNODC provides assistance to states, helping them to draft laws and identifying workable strategies. In 2009, the UNODC launched a worldwide campaign to raise awareness about the extent and nature of human trafficking. Other international agencies involved in combating trafficking and raising awareness include the International Organization for Migration (IOM), the UNHCR, the Office of the High Commissioner for Human Rights (OHCHR), the United Nations Children's Fund (UNICEF), and the United Nations Development Fund for Women (UNIFEM). According to Susan Martin and Amber Callaway (2011, 239), "the international regime to combat human trafficking is in its infancy." The Palermo Protocol is still relatively new and so it will be important to see how well this legislative framework deals with the problem of human trafficking across the globe. At a regional scale, in 2005, the Council of Europe Convention on Action Against Trafficking in Human Beings was developed and signed by 43 member states of the Council of Europe. The convention is aimed at combating human trafficking, assisting and protecting its victims, and ensuring effective prosecution of its perpetrators. In 2000, the United States enacted the Trafficking Victims Protection Act, introducing similar measures to combat human trafficking.

HUMAN SMUGGLING AND BORDER CONTROL

There is a relationship between human smuggling and increased border security. The more difficult it is to move across international borders without proper documentation, the more migrants are forced to utilize smugglers to help them, and many times these migrants incur great costs as a result. Wayne Cornelius (2001), a U.S. scholar on migration and Mexican politics, suggests that increased border patrol and enforcement along the border between Mexico and the United States has increased the costs of illegal entry for many Mexican migrants. He examines U.S. border enforcement strategies since 1993 when border crossing became tougher and suggests that the costs to Mexican migrants doubled, tripled, and, in some cases, quadrupled, depending on the services provided by smugglers and the entry point for the migrant. Coyotes—smugglers who guide migrants across international borders for profit—increasingly charge higher fees. For instance, in parts of Arizona, the fee to be smuggled to Phoenix was US$150 in 1999 and had jumped to between US$800 and US$1,300 by the summer of 2000 (Cornelius, 2001). Migrants typically obtain the money to pay coyote fees by borrowing from family members and, as Koser (2007) suggests, migrants and their families become indebted to these people for increasingly large sums of money.

Cornelius' work stands in contrast to work by Koser (2007). He suggests that the overall cost for smuggling migrants has declined in recent years, likely due to increasing competition in the smuggling business. Koser (2004) interviewed migrant smugglers in Afghanistan and Pakistan and found that the amount they charged for smuggling had remained the same but that the payment process had changed. Migrants tend not to pay smugglers up front, fearing the smuggler will disappear with the money and never help them reach their destination. As a result, smugglers typically asked for a deposit and then the remaining payment upon arrival. Once they arrived in the destination country, however, some migrants were exploited by smugglers to whom they were now indebted. In more recent years, smugglers receive payment through a third party; payment is made in full but to the third party who then passes the money along to the smuggler once the migrant arrives at his or her destination.

Aside from the financial costs incurred by migrants, human smuggling can also be dangerous. The biggest danger lies in the means of transportation, which are sometimes unsafe. Migrants have drowned at sea, or suffocated in sealed containers or in the back of trucks. Smugglers have been known to rape and abuse female migrants while in transit.

CHAPTER SUMMARY

In this chapter, we examined a number of important forces influencing migration policy. First, we examined the importance of labor needs in the evolution of migration legislation in different countries. At different times in their history, developed nations have encouraged labor migration. In more recent decades, migration policy has focused on encouraging high-skilled migration but has struggled to control low-skilled irregular migration.

Second, we identified irregular migration as a growing concern among developed nations and discussed a number of strategies that have been employed to help control unauthorized migrants. These strategies include the development of employer sanctions and legalization programs.

Third, we noted that migration regulation occurs at different scales. Nations can come together to control and regulate the movement of people. Regional entities such as the European Union have evolved, which has allowed more freedom of movement between member states. In contrast, in the United States, there is evidence that migration policy has devolved to subnational governments as states and local jurisdictions attempt to control migration.

Fourth, we examined the role of displacement in the evolution of migration policy internationally. We focused on the emergence of an international refugee regime that offers guidance, assistance, and protection to refugees across the world. Of great importance is the UNHCR. We examined ways in which the role of UNHCR has changed over time. Its work has become more global in focus, and the agency has moved beyond the mere legal protection of refugees to the provision of humanitarian assistance to refugees and even Internally Displaced People. As part of this discussion, we examined the significant contribution made by nation-states in the provision of asylum. The developed world is confronted with migrants arriving at their borders seeking refuge. These asylum seekers are increasingly denied refugee status and, especially in a climate of concerns about national security, legal restrictions on who receives refugee status are unlikely to be relaxed anytime soon.

Finally, we examined the problems of human trafficking and human smuggling. Both of these problems are difficult to solve. Recent international instruments have been implemented to eliminate human trafficking and prosecute those who engage in the trafficking of humans. The modern-day slave trade, while condemned by international law, is a substantial industry (Kaye and McQuade, 2007). It subjects its victims to insecure, degrading work, similar to past forms of slavery, and exists in different forms including debt bondage, forced labor, and human trafficking. According to the International Labor Organization (ILO), there are at a minimum an estimated 12.3 million people living in forced labor in Africa, Europe, Asia, the Americas, and the Middle East. The ILO's *2002 Global Report* on child labor estimates that there were 8.4 million children in slavery worldwide. Organizations have raised awareness of this problem, but more needs to be done to protect humans from various forms of modern-day slavery and exploitation.

GUIDE TO FURTHER READING

There are some excellent online sources related specifically to refugees. We refer the reader to www.forcedmigration.org for research and multimedia resources related to human displacement. We also refer the reader to the website of the United Nations High Commissioner for Refugees (www.unhcr.org) and that of the U.S. Committee for Refugees and Immigrants (www.refugees.org). In the United States, the American Civil Liberties Union (www.aclu.org) provides in-depth resources on the legal, political, and public policy dimensions of migration issues.

MIGRATION AND THE FUTURE

Urbanization, human rights, terrorism, national security, ethnic conflict, economic growth, and environmental sustainability are big topics all connected in some way to global migration. In this book, we have explored the multiple dimensions of these connections. In particular, we have identified: how migrants tend to move to cities; how helping refugees is a question of human rights; how terrorism, national security, and ethnic conflict affect migration policy and societal relations; how migrants influence the economy; and, in this chapter, the nexus between migration and the environment. In the words of Stephen Castles and Mark Miller (2009, 299),

> international migration has never been as pervasive, or as socio-economically and politically significant, as it is today. Never before have political leaders accorded such priority to migration concerns. Never before has international migration seemed so pertinent to national security and so connected to conflict and disorder on a global scale.

This book has sought to gain a succinct and meaningful understanding of the main tenets of migration across the globe by focusing on the history and geography of migration, its impacts on the economy and society, and the political and policy responses to global migration.

In this final chapter, we provide a concluding summary of the chapters in this book. But we begin with a discussion of three aspects of global migration that we believe will be very important in the future. First, we explore the intersection between migration and the environment. We believe that in the future there will be increasing movements of people as a result of environmental disasters, rising sea levels, and overall climate change. Second, global migration will matter more in the future because of the emergence and then persistence of what is referred to as lifestyle migration. Relatively affluent people will continue to migrate to other countries for reasons of lifestyle. As globalization increasingly impacts residents of the developing world, we believe that migration will become more prevalent and an important feature of the political and policy debates worldwide.

The final aspect we examine is the continued challenge of coping with irregular low-skilled migration. Concerns over national security and threats posed by international terrorism are likely to encourage ever-more-stringent measures to monitor the movement of people into more advanced economies; and yet, at the same time, the polarization between nations is likely to continue to encourage the migration of low-skilled workers from developing countries to nations of the developed world. When considered together, these two processes will lead to even larger numbers of irregular migrants.

MIGRATION AND THE ENVIRONMENT

Throughout human history, people have migrated as a response to changes in the natural and built environment. Humans have moved from place to place in search of food provided by nature around them. People have always been aware of the need to migrate because of shifting seasons or preferences for certain climatic conditions. The connection between migration and the environment is certainly not a new or recent occurrence. Throughout history, large-scale climatic events have caused the forced migration of many people across the globe. However, in recent decades, migration as it relates to environmental change has become of growing concern in light of our understanding about problems associated with climate change and sea-level rise.

Climate change refers to major changes in the Earth's temperature, rainfall, snowfall, and other weather events for extended periods of time. There can be natural processes, such as changes in the Sun's intensity or the Earth's orbit, that can cause climate to change, but there is now indisputable evidence that human activities such as the burning of fossil fuels, deforestation, and urbanization are profoundly affecting the Earth's climate (Solomon *et al.*, 2007). Increased greenhouse gases such as carbon dioxide, methane, and nitrous oxide are the consequences of these human activities, and carbon dioxide emissions are the primary environmental concern. As a result of heat-trapping gases such as carbon dioxide, temperatures across the globe have climbed, causing major shifts in the Earth's climate equilibrium (Dessler and Parson, 2010).

In April 2010, the U.S. Environmental Protection Agency (EPA), with the aid of many scientists, produced a report, *Climate Change Indicators in the United States*, to demonstrate, using some basic markers, the main components of climate change in the United States. The report emphasizes problems such as the increased frequency and intensity of extreme climate events such as drought, heat waves, and tropical storms. With the overall warming of the planet, heat waves are expected to continue; the rise in average temperatures "speeds up" the Earth's water cycle, increasing evaporation such that there will be more water available for precipitation in some areas while other areas will experience drought and, as a consequence, in some cases, desertification (U.S. EPA, 2010). Also, climate change contributes to sea-level rise in two ways. First, as the oceans warm, they expand, resulting in a rise in sea level. Second, with the warming, glaciers and ice sheets melt faster, which causes rising sea levels. As the seas rise, villages, towns, and cities that are located along coastlines become extremely vulnerable—being slowly engulfed by incoming, rising waters. Also, with sea-level rise, storm surges from tropical storms and hurricanes are potentially more severe.

Environmental events and processes associated with global climate change are predicted to greatly increase human migration both internally within nations and across international borders. The nature and extent of the effect, however, is difficult to measure. For instance, Norman Myers (2002) suggests that large-scale environmental change will displace some 50 million to 250 million people by 2050. While some experts disagree with the figures offered by Myers (Biermann and Boas, 2007; McAdam, 2011), the *Stern Review of the Economics of Climate Change* in 2007 (Stern, 2007) suggests that estimates of 200 million displaced by climate change are rather conservative. Still others suggest that alarmist predictions, typically aimed at publicizing the climate change problem, lead to stigmatization of migrants, especially those from less developed regions of the world (Castles, 2011).

Part of the difficulty with accurately measuring human displacement caused by environmental events relates more specifically to the resilience of people and communities (Flynn, 2007). People adapt to their natural surroundings. Some flee when there is natural disaster, never to return. Others stay, and ride out the storm. Others leave temporarily. Some communities adapt slowly and plan accurately the effects of climatic problems as sea levels rise. Others do not have the resources to effectively manage environmental change or recover from an environmental disaster. Predicting accurately what people can and will do in an environmental event is difficult. For this reason, measuring exact numbers of people who will be temporarily or permanently displaced and migrate elsewhere is not easy, and yet most experts agree that the effects of climate change are likely to provoke some degree of displacement (McAdam, 2011).

People who move because of environmental events are termed environmental migrants. The International Organization for Migration (2007, 6) defines environmental migrants as:

> persons or groups of persons who, for compelling reasons of sudden or progressive changes in the environment that adversely affect their lives or living conditions, are obliged to leave their habitual homes, or choose to do so, either temporarily or permanently, and who move either within their country or abroad.

Within this definition, a distinction is made between migration resulting from catastrophic events (e.g. hurricane) and migration resulting from the slow onset of gradual environmental change (e.g. desertification). In the latter case, people may adapt over long periods of time, making it difficult to determine if environmental change is the only factor causing migration.

In fact, among scholars, there remains disagreement over whether climatic events and climate change are the most important causes of population movement and displacement. Decisions to move after a catastrophic event or as a result of slow-moving environmental processes are affected by a number of forces. These include the overall socio-economic conditions and political structures of the communities impacted. Ruth Haug (2002), in a study of the effects of severe drought on Hawaweer in northern Sudan, found that some displaced Hawaweer returned because new livelihood opportunities were established. Specifically, displaced people who returned did so to reclaim a farm or get a job, and in some cases they secured a technical job operating pumps, tractors, grinding mills, and other equipment needed to adapt to the possibility of future droughts. Others returned because they could not find employment in the places to which they temporarily migrated. As Stephen Castles (2002) argues, the underlying causes for displacement include socio-political and socio-economic factors. There are multiple drivers to human migration, and the ability of people and communities to adapt to climate change and any associated catastrophic or slow-moving environmental events depends on a multitude of conditions related to social, economic, and political structures. How we cope with environmental change will be important for future global migration.

SUPERSTORM SANDY AND DISPLACEMENT

The United States witnessed the effects of an extreme storm event when Hurricane Sandy pounded New York, New Jersey, Massachusetts, Maryland, and other states along the eastern coastline on October 29, 2012. Much of the areas hit hard by Superstorm Sandy were unprepared for the storm surge and its impact. Much of Lower Manhattan was without electrical power for nearly a week. The New York City subway system was shut down for several days, and thousands of residents, particularly in New York and New Jersey, lost their homes.

On October 31, 2012, New York Governor Andrew Cuomo held a press briefing about the effects of Hurricane Sandy and stated, "It's a longer conversation, but I think part of learning from this is the recognition that climate change is a reality. Extreme weather is a reality.

Figure 6.1 Aerial views of the damage caused by Hurricane Sandy to the New Jersey coast.
Source: New Jersey National Guard, October 31, 2012. Available online at http://commons.wikimedia.org/wiki/File%3A121030-F-AL508-081c_Aerial_views_during_an_Army_search_and_rescue_mission_show_damage_from_ Hurricane_Sandy_to_the_New_Jersey_coast%2C_Oct._30%2C_2012.jpg (accessed May 12, 2013).

It is a reality that we are vulnerable" (Lovett, 2012). There is some debate over the exact relationship between climate change and Superstorm Sandy, but many scientists agree that warmer oceans contributed to the severity of the storm, and sea-level rise and sinking coastlines likely contributed to the devastating effects. The National Hurricane Center attributed 72 deaths directly to Sandy and 87 others indirectly to the storm, from such problems as hypothermia resulting from power outages and accidents from the cleanup efforts. The total number of dead was 159.

According to Governor Chris Christie of New Jersey, six months after the storm hit there were still an estimated 39,000 New Jersey families displaced, some residents were still fighting with insurance companies to salvage something from their devastated homes, and others were unable to find or afford other rental properties. Five months after the storm, some 2,000 displaced people remained in hotels in New York City (Navarro, 2013). Many of these families and individuals are lower-income households. Even in the United States, with the potential availability of massive resources, displacement from violent climatic events occurs and can be long-term, especially for lower-income individuals.

LIFESTYLE MIGRATION

In recent years, there has been increased interest in understanding the motivations of those who migrate for reasons of lifestyle. Many who migrate for such reasons are relatively affluent. In general, the migration of affluent individuals is little studied in the literature on global migration (Benson and O'Reilly, 2009a). There are some exceptions, as we explored in Chapter 4, which include studies on the mobility of a professional class of transnational elites within the international migratory system (Beaverstock, 2005). In addition, there are studies on international retirement migration (King, Warnes, and Williams, 1998). According to some scholars, the trend of migration by a relatively affluent group of people that seek a better quality of life is growing and is significant (Benson and O'Reilly, 2009a). In this section, we explore lifestyle migration and suggest that it will be a critical feature of future global migration.

In a recent book, *Lifestyle Migration*, Michaela Benson and Karen O'Reilly describe the tendency to migrate for specific reasons of lifestyle as non-typical of migration movements. They state (2009a, 1):

> This way of life can be distinguished from that sought by other migrants, such as labour migrants, refugees and asylum-seekers, in its emphasis on lifestyle choices specific to individuals of the developed world; migration for these migrants is often an antimodern, escapist, self-realization project, a search for the intangible "good life."

The various chapters in *Lifestyle Migration* include examples of migration by affluent people to places such as Varanasi, India, to the South of France, Florence, Italy, and Didim, Turkey. The authors attempt to uncover the motivations behind these movements. For many, migration is motivated by a desire to acquire a new way of life that holds more meaning. Lifestyle migrants seek to escape the individualism, materialism, and monotony of contemporary life. They seek a new sense of community, and they desire a sort of refuge from what Benson and O'Reilly (2009a, 4) describe as "the excessive consumption, stifling working practices, unpalatable futures and daily misery they have willingly fled." Lifestyle migrants seek closeness with nature, the "good vibes of India," or the mythical and imagined landscape of the Grand Traverse in rural Michigan. In this context, the rural is seen as restorative, and the rural lifestyle is imagined as offering an opportunity for self-renewal. In fact, motivating forces for lifestyle migrants suggest a desire for self-realization, and these more affluent migrants often see themselves as pioneers in their quest for a new type of lifestyle that is less individualistic in nature (Benson and O'Reilly, 2009a).

Unfortunately, as described by Benson and O'Reilly (2009a), lifestyle migrants, while seeking a renewed sense of community, often have little contact with the local people who live in their new destination society. Lifestyle migrants tend to socialize with other lifestyle migrants in the area rather than establish any deep ties with the native population. There is often a contradiction between the aspirations of lifestyle migrants and the reality of their lives. Like other migrants, they experience language barriers. They are often rejected or slighted by the local residents. Sometimes they find grave cultural and social differences between themselves and the native population that enhance their feelings of alienation.

Lifestyle migration tends to occur in particular geographies. Benson and O'Reilly (2009b) identify four categories of place. They include residential tourist destinations along the coast. Examples include such places as the Algarve, Malta, and the Costa del Sol for migrants that seek the Mediterranean lifestyle. Other places are rural; places Benson and O'Reilly (2009b, 6) describe as imagined locations that "offer migrants a sense of stepping back in time, getting back to the land, the simple or good life, as well as a sense of community spirit." Examples include the migration of British people to France. Other destinations for lifestyle migrants include more bohemian places that they (2009b, 6) suggest are "characterised by certain spiritual, artistic, or creative aspirations and unique 'cultural' experience." Examples include Deia in Mallorca and Mykonos in Greece. In this case, lifestyle migrants are seeking an alternative lifestyle, away from mainstream society.

Thus far, we have described the individual motivations, consumptive patterns, and realities of lifestyle migration, but there are also larger macro-forces that add to our understanding of this process. By and large, lifestyle migrants are more affluent and have the economic privilege to migrate for lifestyle reasons. Many lifestyle migrants own two homes: they live in one part of the world for some part of the year; and in another place in the other part of the year. Other lifestyle migrants travel extensively between these places to visit family and friends that they left behind. Indeed, they can afford to do so. In the case of lifestyle migrants within Europe, many people take advantage of the public policy environment that allows Europeans to migrant internally throughout the continent.

Lifestyle migration is more possible today than previously because of newer, more efficient, and cheaper modes of communication and transportation. Online networks for exchanging information transnationally are pertinent to learning about new destinations. Migrants have the ability to travel from their original homes to new destinations more easily now than ever before. In the future, modes of communication and transportation have the potential to improve, making it more possible for lifestyle migrants to have transnational living arrangements. More important, living standards among the more affluent of developed nations are rising so more people "can make informed and financially viable choices about their lifestyles" (Benson and O'Reilly, 2009b, 12). At the same time, employment structures allow for increasing flexibility in work life, especially among the more affluent professional class. Then, of course, there are those agents on the production side of lifestyle migration—the real estate agents, property management companies, financial institutions, and, increasingly, government policies—that enable and encourage such migration (Benson and O'Reilly, 2009b). We suspect that lifestyle migration will increase in the future as the more affluent of the developed world seek new forms of consumption in an increasingly integrated and connected global world.

IRREGULAR MIGRATION CONTINUES

At the other end of the socio-economic spectrum, the irregular migration of poor and low-skilled workers will continue to be an important feature of the global movement of people in the future. As we explored in Chapter 5, especially since the 1980s, governments of more advanced economies have introduced public policies aimed at controlling irregular migration. Yet, despite these efforts, people continue to migrate across international borders without legal documentation. Asylum seekers, upon refusal of refugee status, remain in destination societies illegally. Families of undocumented migrants make efforts to join them. Irregular migration, particularly of poorly skilled migrants, remains a challenge for policymakers.

Part of the difficulty in effectively controlling migration lies in the inherent forces of globalization and the needs of the global economy. As Stephen Castles and Mark Miller (2009, 306) suggest, "In an increasingly international economy, it is difficult to open borders for movements of information, commodities and capital and yet close them to people." In many respects, globalization makes it easier to migrate. Through the global dispersion of information, migrants can find out about destination countries more easily now than in the past (Ritzer, 2011). An industry of human smugglers enables poorly skilled migrants to cross international borders illegally.

Most importantly, the global economy has specific labor needs. On the one hand, there is a need for high-skilled technicians and professionals in the areas of finance, information technology, and advanced producer services. Migration policy tends to favor the in-migration of these types of workers. On the other hand, the global economy also requires low-skilled workers—such as domestic servants, taxi-drivers, janitors, gardeners, childcare workers, hairdressers, waitresses, food service staff, and security guards. As we explored in our discussion of global cities in Chapter 4, cities with a concentration of high-skilled workers tend also to be places that attract low-skilled workers—low-skilled service workers are necessary to cater to the needs of people in high-skilled employment. Many native residents are unwilling to accept menial jobs, at least for long periods of time. Thus, foreign workers are needed, and yet migration policy tends to seek to prevent the in-migration of low-skilled workers, many of whom become irregular migrants. Migration policy does not reflect the needs of the global economy in the twenty-first century. Some suggest that governments need to reduce the number of policies aimed at restricting the movement of people (United Nations Programme, 2009). Many nations are faced with this policy paradox, and it will likely continue to be the case in the future.

For many citizens and politicians of destination countries, irregular migration means "unwelcome migration." Irregular or undocumented migration has emerged in the public discourse to signify a threat to national identity and national security (Vicino, 2013). Let us first discuss the migrant as a threat to national identity. As we explored in Chapter 3, the influx of large numbers of migrants into another nation can lead to conflict between newcomers and natives. Nowadays, many migrants to developed nations migrate from underdeveloped nations. They have different customs, religions, languages, and skin color to the native population. The poorly skilled irregular migrant has come to symbolize the nation's inability to control core large-scale social transformations occurring within its borders. There is concern that poorly skilled migrants do not assimilate into the native society and adopt its customs, language, and culture, but rather change the native society in profound ways. Yet, life is not easy for low-skilled irregular migrants. These are people who are often marginalized economically within destination societies, separated from family members, and feel isolated and threatened by law enforcement officials and border restrictions. Acceptance of irregular migrants is increasingly difficult as tensions persist, and irregular migrants become more segregated and detached from the destination society.

The irregular migrant has also become a figurative threat to national security. Concern over terrorism, particularly in the United States and Europe, has served to reinforce the need for restrictions on migration. Anyone flying today compared with even 10 or 15 years ago recognizes the enhanced security measures that have been put in place to clear border control in U.S. and European airports. Migration restrictions for students from countries in Asia in particular who wish to study in Europe and the United States have increased over time. Many of these measures took effect after the attacks on the World Trade Center and the Pentagon on September 11, 2001. In Chapter 3, we explored the impact of recent terrorist attacks on societal perceptions of migration and migrants. Public opinion and public policies have leaned more toward stricter migration control as a result. Previously, there was tacit government consent to allow migrants to enter illegally, especially in the United States (Castles and Miller, 2009). This, we believe, is changing as irregular migration becomes increasingly tied to issues of national security and foreign policy. Tightening borders and restricting migration in the interest of national security will lead to more people entering under difficult circumstances as irregular migrants.

SUMMARY OF THE BOOK

This book examines the history and geography of global migration, its social implications, the relationship between migration and the global economy, and the current and past public policy environments as they relate to the regulation and control of the movement of people. In Chapter 1, we provide the reader with key definitions related to migration. We examine the various theories around why migration happens.

Some theorists focus on the contribution that individuals and communities make as part of the migration decision-making process. People, their families, and communities weigh the costs and benefits of migration, and, based on the results, they decide whether or not to move. For other theorists, it is large structural forces that shape migration processes. For some, the structure of the labor market matters. Poorly paid low-skilled jobs are unattractive to native workers, and so businesses, institutions, and governments must recruit foreign workers. For others, larger processes related to the capitalist mode of production create situations where people in underdeveloped countries become increasingly mobile and must move to find employment. Some move internally, especially from rural to urban areas. Others move internationally, mostly to cities of the more advanced economies in the North, but increasingly they move to oil-rich nations in the Persian Gulf and other more advanced economies in the Global South.

As we demonstrate in Chapter 2, global migration has a long history, playing a key role in our understanding of the impacts of colonialism, the nature of industrial capitalist expansion, and the character of the current postindustrial global economy. During the colonial period, there was a large outflow of population from Europe to parts of the colonized world. There was also the forced migration of slaves from continental Africa to the Americas, and indentured servants from Europe and, in latter periods, Asia. Migration during the period of industrialization included the movement of large numbers of Europeans to the Americas, a shifting of people internally within Europe, and the mass migration of people from India and China to countries in Southeast Asia. These population movements were influenced by integration of an increasingly global economy centered on the labor needs of rising industrial production in Europe and North America. The influence of globalization continued during the current postindustrial period. Migration continued in "classical immigration" countries. This time, migration patterns incorporated the movement of Latin Americans and Asians into the United States, and Asians into Australia. In addition, Europe experienced the in-migration of minority populations, especially from Africa and the Middle East. Finally, there has been a recent movement of people into the oil-rich Persian Gulf nations.

As we explore in Chapter 3, global migration has important social implications. Migrants can change destination societies. Often, migrants develop new ethnic communities. This can revitalize old neighborhoods and introduce new cuisine and culture to destination societies. Sometimes migrants incorporate effectively into their new society. Governments, through policies and rhetoric of multiculturalism, aim to demonstrate the societal benefits of new migration. In some cases, migrants do not assimilate, but rather act as transmigrants, living comfortably in two separate cultures. In recent years, especially with terrible acts of terrorism and a lack of migrant integration, there has been an anti-immigrant backlash in some destination societies. We explore these effects and consider the relationship between migration and gender. Migration has always included the movement of women but, in more recent years, women are migrating for economic reasons. Female migrants often face discrimination in the labor market and find themselves in poorly paid gender-segregated occupations. They also are often victims of trafficking and abuse. Discrimination also exists for lesbian, gay, bisexual, and transgendered (LGBT) migrants. The migration policies of many countries still discriminate on the basis of sexual orientation and sexual identity and many migrants move from countries where LGBT people are persecuted.

As we examine in Chapter 4, migration is important to the global economy. We look at the impacts of migration on the wages and employment among workers of immigrant-receiving nations. Low-skilled workers can often be impacted by the inflow of migrants. The impact of migrants on the domestic workforce can be a strong impetus for migration control. Also, migrants tend to be particularly concentrated in cities central to the global market place. We examine the migration of highly skilled workers to cities, some moving from less developed to more developed nations. This can lead to brain drain among developing countries, although in some cases it can lead to brain gain as migrants return to their country of origin with particular skills and education that, in the right circumstances, provide their home country with economic benefits. Scholars and the international development community have begun to explore the connections between migration and development in less developed nations, and, in some cases, raise new hopes that remittances can greatly benefit developing societies.

As we explore in Chapter 5, public policies on migration have evolved over time. In the past, there was little restriction of the movement of people from nation to nation, but with emergence of nation-states and the integration of a global economy and market-place, nation-states have a more vested interest in controlling migration across jurisdictional boundaries. There are a growing number of irregular migrants worldwide, and the mass movement of refugees forced to migrate because of conflict, environmental disasters, and persecution is an international issue. As globalization persists, there is a tendency to allow the free flow of capital, goods, services, and information, but the opposite generally occurs with the movement of people globally. Rather, nation-states seek to control human migration across their borders, and there is political tension and strife around the issue of immigration in many countries. In the future, it will become increasingly necessary to develop an adequate legal framework for global migrants who enter or stay in a country illegally or because of some sort of conflict, disaster, or persecution in their own country.

In sum, global migration is a complex process that touches all facets—social, economic, political, and cultural—of a migrant's life as well as the new societies that they call home. There is a long history of migration that dates back to the beginning of human civilization. Migration touches all geographies, and people have migrated to and from all parts of the world. They have brought with them traditions, customs, and new ways of life. Time and again, societies and their economies have been fundamentally transformed as new people depart and arrive. At times, these changes have presented political and policy challenges, and institutions have responded in a variety of ways. Thus, to understand "the basics" of global migration requires a balanced approach that considers the politics, economics, and humanistic impacts that shape how and why we move. Indeed, we live in a smaller, more connected world and our lives will undoubtedly be impacted by global migration.

GUIDE TO FURTHER READING

For an excellent resource for the study of the relationship between global migration and the environment see, Laczko, F. and Aghazarm, C. (Eds) (2009). *Migration, Environment and Climate Change: Assessing the Evidence*. Geneva, Switzerland: International Organization for Migration.

For an excellent resource on the study of lifestyle migration see, Benson, M. and O'Reilly, K. (Eds) (2009). *Lifestyle Migration: Expectations, Aspirations and Experiences*. Surrey, UK: Ashgate Publishers.

REFERENCES

Adams, R.H., Jr. and Page, J. (2003) International Migration, Remittances, and Poverty in Developing Countries. Policy Research Working Paper Series 3179, The World Bank.

Adelman, J. (1995) European Migration to Argentina, 1880–1930. In Cohen, R. (Ed.) *Cambridge Survey of World Migration*. Cambridge, UK: Cambridge University Press, pp. 215–19.

Alba, R.D. and Nee, V. (1997) Rethinking Assimilation Theory for a New Era of Immigration. *International Migration Review* 31(4): 826–74.

Ali, S. (2010) *Dubai: Gilded Cage*. New Haven, CT: Yale University Press.

Associated Press (2006) Pa. Mayor Tells Illegal Immigrants to Go. July 14. Available online at http://usatoday30.usatoday.com/news/nation/2006-07-13-hazleton-immigrants_x.htm (accessed June 14, 2013).

Australian Government (2013) *Fact Sheet 8 – Abolition of the "White Australia" Policy*. Department of Immigration and Citizenship. Available online at http://www.immi.gov.au/media/fact-sheets/08abolition.htm (accessed May 25, 2013).

Ayres, R. and Barber, T. (2006) Statistical Analysis of Female Migration and Labor Market Integration in the EU. Integration of Female Immigrants in Labour Market and Society Working Paper 3. Oxford, UK: Oxford Brookes University.

Bade, K. (2003) *Migration in European History*. Malden, MA: Blackwell Publishing.

Bade, K.J. (1995) Germany: Migrations in Europe up to the End of the Weimar Republic. In Cohen, R. (Ed.) *Cambridge Survey of World Migration*. Cambridge, UK: Cambridge University Press, pp. 131–5.

Baily, S.L. and Miguez, E.J. (Eds) (2003) *Mass Migration to Modern Latin America*. Lanham, MD: Rowman & Littlefield Publishers.

Banting, K. and Kymlicka, W. (2006) *Multiculturalism and the Welfare State: Recognition and Redistribution in Contemporary Democracies*. New York: Oxford University Press.

Barth. F. (1969) *Ethnic Groups and Ethnic Boundaries*. Long Grove, IL: Waveland Press.

Beaverstock, J.V. (2005) Transnational Elites in the City: British Highly-skilled Inter-company Transferees in New York City's Financial District. *Journal of Ethnic and Migration Studies* 31(2): 245–68.

Belot, M. and Hatton, T. (2008) Immigrant Selection in the OECD (February 2008). CEPR Discussion Paper No. DP6675. Available online at: http://ssrn.com/abstract=1140957 (accessed May 16, 2013).

Belser, P., De Cock, M. and Mehran, F. (2005) *ILO Minimum Estimate of Forced Migration in the World*. Geneva: International Labor Organization.

Benson, M. and O'Reilly, K. (Eds) (2009a) *Lifestyle Migration: Expectations, Aspirations and Experiences*. Surrey, UK: Ashgate Publishers.

——(2009b) Migration and the Search For A Better Way of Life: A Critical Exploration of Lifestyle Migration. *The Sociological Review* 57(4): 608–25.

Betts, A. (Ed.) (2011) *Global Migration Governance*. Oxford: Oxford University Press.

Biermann, F. and Boas, I. (2007) Preparing for a Warmer World: Towards a Global Governance System to Protect Climate Refugees. Global Governance Working Paper No. 33, November 2007.

Blackburn, R. (2010) *The Making of New World Slavery: From the Baroque to the Modern, 1492–1800, Second Edition*. London: Verso.

——(2011) *The American Crucible: Slavery, Emancipation and Human Rights*. London and New York: Verso.

Borjas, G. (2001) *Heaven's Door: Immigrant Policy and the American Economy*. Princeton, NJ: Princeton University Press.

——(2004) *Increasing the Supply of Labor Through Immigration: Measuring the Impact on Native-born Workers*. Washington, DC: Center for Immigration Studies.

Borjas, G.J. (1989) Economic Theory and International Migration. *International Migration Review* 23(3): 457–85.

Bourbeau, P. (2011) *The Securitization of Migration: A Study of Movement and Order*. Oxford, UK: Routledge.

Boyd, M. (1989) Family and Personal Networks in International Migration: Recent Developments and New Agendas. *International Migration Review* 23(3): 638–70.

Boyd, M. and Grieco, E. (2003) *Women and Migration: Incorporating Gender into International Migration Theory*. Migration Policy Institute. Available online at http://www.migrationinformation.org/Feature/display.cfm?ID=106 (accessed May 14, 2013).

Brandt, L. (1985) Chinese Agriculture and the International Economy 1870–1913: A Reassessment. *Explorations in Economic History* 22: 168–80.

Brettell, C.B. and Hollifield, J.F. (2007) *Migration Theory: Talking Across Disciplines*. New York: Routledge.

Calavita, K. (1996) The New Politics of Immigration: "Balanced-Budget Conservatism" and Prop. 187. *Social Problems* 43: 284–305.

Camarota, S. (2007) *Immigrants in the United States: A Profile of America's Foreign-born*. Washington, DC: Center for Immigration Studies.

Canaday, M. (2009) Thinking Sex in the Transnational Turn: An Introduction. *The American Historical Review* 114(5): 1250–57.

Castles, S. (2002) Environmental Change and Forced Migration: Making Sense of the Debate, New Issues in Refugee Research. Working paper No. 70. Geneva, Switzerland: United Nations High Commissioner for Refugees.

——(2004) The Factors that Make and Unmake Migration Policies. *International Migration Review* 38(3): 852–84.

——(2011) Concluding Remarks on the Climate Change–Migration Nexus. In Piguet, E., Pécoud, A. and De Guchteneire, P. (Eds) *Migration and Climate Change*. Cambridge, UK: Cambridge University Press, pp. 415–27.

Castles, S. and Miller, M.J. (2009) *The Age of Migration, Fourth Edition*. New York: Palgrave Macmillan.

Chan, E. (2013) Ethnic Enclaves and Niches: Theory. In Ness, I. and Bellwood, P. (Eds) *The Encyclopedia of Global Human Migration*. Hoboken, NJ: Wiley-Blackwell.

Chiswick, B. and Hatton, T. (2003) International Migration and the Integration of Labor Markets. In Bordo, M.D., Taylor, A.M. and Williamson, J.G. (Eds) *Globalization in Historical Perspective*. Chicago, IL: University of Chicago Press, pp. 65–120.

Chiswick, B.R. and Miller, P.W. (Eds) (2012) *Recent Developments in the Economics of International Migration*. Northampton, MA: Edward Elgar Publishing.

Cohen, R. (1995) *Cambridge Survey of World Migration*. Cambridge, UK: Cambridge University Press.

Coleman, M. (2007) Immigration Geopolitics Beyond the Mexico–US Border. *Antipode* 39(1): 54–76.

——(2008) US Immigration Law and its Geographies of Social Control: Lessons from Homosexual Exclusion During the Cold War. *Environment and Planning D: Society and Space* 26(6): 1096–114.

Comaroff, J.L. and Comaroff, J. (2009) *Ethnicity, Inc.* Chicago, IL: University of Chicago Press.

Cornelius, W.A. (2001) Death at the Border: Efficacy and Unintended Consequences of US Immigration Control Policy. *Population and Development Review* 27(4): 661–85.

Coulombe, S. and Tremblay, J.F. (2009) Migration and Skills Disparities across the Canadian Provinces. *Regional Studies* 43(1): 5–18.

Couton, P. (2013) Ethnic and National Pride. In Ness, I. and Bellwood, P. (Eds) *The Encyclopedia of Global Human Migration*. Hoboken, NJ: Wiley-Blackwell. Available online at http://onlinelibrary.wiley.com/doi/10.1002/9781444351071.wbeghm198/abstract (accessed May 5, 2013).

Crampton, T. (2005) Behind the Furor, the Last Moments of Two Youths. *New York Times*, November 7.

Cullen, P. (2000) *Refugees and Asylum-Seekers in Ireland*. Cork, Ireland: Cork University Press.

Daniels, R. (2002) *Coming to America: A History of Immigration and Ethnicity in American Life, 2nd Edition*. New York: Harper Perennial.

De Jong, G.F. (2002) Expectations, Gender, and Norms in Migration Decision-Making. *Population Studies* 54: 307–19.

Dessler, A. and Parson, E.A. (2010) *The Science and Politics of Global Climate Change: A Guide to the Debate, Second Edition*. Cambridge, UK: Cambridge University Press.

Docquier, F. and Rapoport, H. (2011) Globalization, Brain Drain and Development. Discussion Paper Series, Forschungsinstitut zur Zukunft der Arbeit, No. 5590.

Docquier, F., Lowell, B.L. and Marfouk, A. (2009) A Gendered Assessment of the Brain Drain. *Population and Development Review* 35(2): 297–321.

Düvell, F. (2011) Irregular Migration. In Betts, A. (Ed.) *Global Migration Governance*. Oxford, UK: Oxford University Press, pp. 78–108.

Ellis, M. (2006) Unsettling Immigrant Geographies: US Immigration and the Politics of Scale. *Tijdschrift voor economische en sociale geografie* 97(1): 49–58.

Engerman, S.L. (1986) Servants to Slaves to Servants: Contract Labour and European Expansion. In Emmer, P.C. (Ed.) *Colonialism and Migration; Indentured Labour Before and After Slavery*. Leiden: Martinus Nijhoff, pp. 263–94.

Erie, S.P. (2006) *Globalizing L.A.: Trade, Infrastructure, and Regional Development*. Stanford, CA: Stanford University Press.

Esbenshade, J.L. (2007) *Division and Dislocation: Regulating Immigration Through Local Housing Ordinances*. Washington, DC: Immigration Policy Center, American Immigration Law Foundation.

Fishman, R. (1987) *Bourgeois Utopias: The Rise and Fall of Suburbia*. New York: Basic Books.

Flynn, S. (2007) *The Edge of Disaster: Rebuilding a Resilient Nation*. New York: Random House.

Fong, E. and Luk, C. (Eds) (2006) *Chinese Ethnic Business: Global and Local Perspectives*. New York: Routledge.

Friedmann, J. (1986) The World City Hypothesis. *Development and Change* 17(1): 69–83.

Gabaccia, D.R. (2013) Gender and Migration. In Ness, I. and Bellwood, P. (Eds) *The Encyclopedia of Global Human Migration*. Hoboken, NJ: Wiley. Accessed online June 16, 2013.

Gallagher, A. (2002) Trafficking, Smuggling and Human Rights: Tricks and Treaties. *Forced Migration Review* 12: 25–8.

Gans, H. (1979) Symbolic Ethnicity: The Future of Ethnic Groups and Cultures in America. *Ethnic and Racial Studies* 2(1): 1–20.

——(1992) Comment: Ethnic Invention and Acculturation, a Bumpy Line Approach. *Journal of American Ethnic History* 1(2): 42–52.

Geertz, C. (1963) *Old Societies and New States: The Quest for Modernity in Asia and Africa*. New York: The Free Press of Glencoe.

Gilmartin, M. (2008) Migration, Identity and Belonging. *Geography Compass* 2 (6): 1837–52.

Glover, S. (2001) *Migration: An Economic and Social Analysis*. London: Research, Development and Statistics Directorate of the Home Office.

Gordon, M.M. (1964) *Assimilation in American Life*. New York: Oxford University Press.

——(Ed.) (1981) *America as a Multicultural Society*. Philadelphia, PA: American Academy of Political and Social Science.

Graham, S. (Ed.) (2004) *Cities, War, and Terrorism: Towards an Urban Geopolitics*. Malden, MA: Wiley-Blackwell.

Hamilton, K. and Yau, J. (2004) The Global Tug-of-War for Healthcare Professionals. Available online at http://www.migrationinformation.org/feature/display.cfm?ID=271 (accessed May 12, 2013).

Hanlon, B., Short, J.R. and Vicino, T.J. (2009) *Cities and Suburbs: New Metropolitan Realities in the US*. New York: Routledge.

Hannerz, U. (1996) *Transnational Connections*. London: Routledge.

Harnett, H.M. (2007) State and Local Anti-Immigrant Initiatives: Can They Withstand Legal Scrutiny. *Immigration and Nationality Law Review* 29: 661.

Harvey, D. (2007) *A Brief History of Neoliberalism*. Oxford: Oxford University Press.

Harzig, C. and Hoerder, D. (2009) *What is Migration History?* Malden, MA: Polity Press.

Hathaway, J.C. and Neve, R.A. (1997) Making International Refugee Law Relevant Again: A Proposal for Collectivized and Solution-Oriented Protection. *Harvard Human Rights Journal* 10: 115.

Hatton, T.J. and Williamson, J.G. (2005) *Global Migration and the World Economy: Two Centuries of Policy and Performance*. Cambridge, MA: MIT Press.

——(2008) *Global Migration and the World Economy: Two Centuries of Policy and Performance*. Cambridge, MA: The MIT Press.

Haug, R. (2002) Forced Migration, Processes of Return and Livelihood Construction Among Pastoralists in Northern Sudan. *Disasters* 26(1): 70–84.

Heath, A.F. and Cheung, S.Y. (Eds) (2007) *Unequal Chances: Ethnic Minorities in Western Labor Markets*. Oxford, UK: Oxford University Press.

Hernández, K.L. (2006) The Crimes and Consequences of Illegal Immigration: A Cross-border Examination of Operation Wetback, 1943 to 1954. *The Western Historical Quarterly* 37(4): 421–44.

Holmes, C. (1995) Jewish Economic and Refugee Migrations, 1880–1950. In Cohen, R. (Ed.) *Cambridge Survey of World Migration*. Cambridge, UK: Cambridge University Press, pp. 148–53.

Huff, G. and Caggiano, G. (2007) Globalization and Labor Market Integration in Late Nineteenth- and Early Twentieth-Century Asia. *Research in Economic History* 25: 285–347.

Huntington, S.P. (2004) The Hispanic Challenge. *Foreign Policy* March 1. Available online at http://www.foreignpolicy. com/articles/2004/03/01/the_hispanic_challenge (accessed May 16, 2013).

Institut National de la Statistique et des Études Économiques (2012) *Foreigners – Immigrants*. Paris: Institut National de la Statistique et des Études Économiques.

Inter-American Development Bank (2004) *Survey of Mexican and Central American Immigrants in the United States*. Washington, DC: MIF-FOMIN and Bendixen Associates.

International Labor Organization (2002) *A Future Without Child Labour*. Global Report Under the Follow-up to the ILO Declaration on Fundamental Principles and Rights at Work. Report of the Director-General, 2002. Geneva, Switzerland.

International Organization for Migration (2007) *Discussion Note: Migration and the Environment*. 94th Session, Document No. MC/INF/288, November.

Iredale, R. (2001) The Migration of Professionals: Theories and Typologies. *International Migration* 39(5): 7–26.

Jenkins, R. (1997) *Rethinking Ethnicity: Arguments and Explorations*. New York: Sage.

Julca, A. (2013) Remittances, Motivation. In Ness, I. and Bellwood, P. (Eds) *The Encyclopedia of Global Human Migration*. Hoboken, NJ: John Wiley & Sons.

Kane, A. and Leedy, T.H. (Eds) (2013) *African Migrations: Patterns and Perspectives*. Bloomington, IN: Indiana University Press.

Kaye, M. and McQuade, A. (2007) *A Discussion Paper on Poverty, Development and the Elimination of Slavery*. London, UK: Anti Slavery International.

Keaten, J. (2005) Riots Rock Paris Suburbs for 8th Day. *The Associated Press* November 3.

Kerwin, D., Brick, K.and Kilberg, R. (2012) *Unauthorized Immigrants in the United States and Europe: The Use of Legalization/Regularization as a Policy Tool.* Available online at http://www.migrationinformation.org/Feature/display.cfm?ID=892 (accessed January 5, 2013).

Khadria, B. (2013) Brain Drain, Brain Gain, India. In Ness, I. and Bellwood, P. (Eds) *The Encyclopedia of Global Human Migration.* Hoboken, NJ: Wiley-Blackwell. Available online at http://onlinelibrary.wiley.com/doi/10.1002/9781444351071.wbeghm073/abstract (accessed June 5, 2013).

King, R., Warnes, A.M. and Williams, A.M. (1998) International Retirement Migration in Europe. *International Journal of Population Geography* 4(2): 91–111.

Kitroeff, N. (2013) Immigrants Pay Lower Fees to Send Money Home, Helping to Ease Poverty. *New York Times*, April 27.

Klein, H.S. (1995) European and Asian Migration to Brazil. In Cohen, R. (Ed.) *Cambridge Survey of World Migration.* Cambridge, UK: Cambridge University Press, pp. 208–14.

Koser, K. (2007) *International Migration: A Very Short Introduction.* New York: Oxford University Press.

Krane, J. (2009) *City of Gold: Dubai and the Dream of Capitalism.* New York: St. Martin's Press.

Kymlicka, W. (2012) *Multiculturalism: Success, Failure, and the Future.* Washington, DC: Migration Policy Institute.

Laczko, F. and Aghazarm, C. (Eds) (2009) *Migration, Environment and Climate Change: Assessing the Evidence.* Geneva, Switzerland: International Organization for Migration.

Lary, D. (2012) *Chinese Migrations: The Movement of People, Goods, and Ideas over Four Millennia.* Lanham, MD: Rowman & Littlefield Publishers.

Latham, A.J.H. and Neal, L. (1983) The International Market in Rice and Wheat, 1868–1914. *Economic History Review* 36: 260–75.

Li, W. (1998) Anatomy of a New Ethnic Settlement: The Chinese Ethnoburb in Los Angeles. *Urban Studies* 35(3): 479–501.

——(2009) *Ethnoburb: The New Ethnic Community in North America.* Honolulu, HI: University of Hawaii Press.

Li, W. and Dymski, G. (2007) Globally Connected and Locally Embedded Financial Institutions: Analyzing the Chinese Banking Sector. In Fong, E. and Luk, C. (Eds) *Chinese Ethnic Business: Global and Local Perspectives.* New York: Routledge, pp. 35–63.

Loescher, G. (2001) *The UNHCR and World Politics: A Perilous Path.* Oxford: Oxford University Press.

Loescher, G. and Milner, J. (2011) UNHCR and the Global Governance of Refugees. In Betts, A. (Ed.) *Global Migration Governance.* Oxford, UK: Oxford University Press, pp.189–209.

Logan, J.R., Alba, R.D., Dill, M. and Zhou, M. (2000) Ethnic Segmentation in the American Metropolis: Increasing Divergence in Economic Incorporation, 1980–90. *International Migration Review* 34(1): 98–132.

Lovett, K. (2012) Hurricane Sandy Death Toll in NY at 26; Gov. Cuomo Blames Climate Change For Increase in Storms. *Daily News* October 31.

Luibheid, E. and Cantu, L. (Eds) (2005) *Queer Migrations: Sexuality, U.S. Citizenship and Border Crossings*. Minneapolis, MN: University of Minnesota Press.

McAdam, J. (2011) Environmental Migration. In Betts, A. (Ed.) *Global Migration Governance*. Oxford, UK: Oxford University Press, pp. 153–88.

McDonald, F. (Ed.) (1997) *Crime and Law Enforcement in the Global Village*. Cincinnati, OH: Anderson Publishing Company.

MacDonald, J.S. and MacDonald, L.D. (1964) Chain Migration, Ethnic Neighborhood Formation, and Social Networks. *Millbank Memorial Fund Quarterly* 42(1): 82–97.

McKeown, A. (2004) Global Migration, 1846–1940. *Journal of World History* 15: 155–89.

Mai, N. and King, R. (2009) Love, Sexuality and Migration: Mapping the Issue(s). *Mobilities* 4(3): 295–307.

Mandel, R. (2008) *Cosmopolitan Anxieties: Turkish Challenges to Citizenship and Belonging in Germany*. Durham, NC: Duke University Press.

Martin, S. and Callaway, A. (2011) Human Trafficking and Smuggling. In Betts, A. (Ed.) *Global Migration Governance*. Oxford, UK: Oxford University Press, pp. 224–41.

Martin, S.F. (2004) *Women and Migration*. Paper presented at the Consultative Meeting on Migration and Mobility and how this Movement Affects Women, United Nations Division for the Advancement of Women, Malmö, Sweden, December 2003. Available online at http://www.un.org/womenwatch/daw/meetings/consult/CM-Dec03-WP1.pdf (accessed May 14, 2013).

Massey, D., Arango, J., Hugo, G., Kouaouci, A., Pellegrino, A. and Taylor, J. (1993) Theories of International Migration: A Review and Appraisal. *Population and Development Review* 19(3): 431–66.

Massey, D.S. (2003) Patterns and Processes of International Migration in the 21st Century. Paper presented at the Conference on African Migration in Comparative Perspective, Johannesburg, South Africa, June 4–7, 2003.

Milanovic, B. (2007) Globalization and Inequality. In Held, D. and Kaya, A. (Eds) *Global Inequality: Patterns and Explanations*. Cambridge, UK: Polity, pp. 26–49.

Miller, M. (2002) Continuity and Change in Postwar French Legalization Policy. In Messina, A. (Ed.) *West European Immigration and Immigrant Policy in The New Century*. Westport, CT: Praeger Publishers, pp. 13–32.

Moch, L. (1995) Moving Europeans: Historical Migration Practices in Western Europe. In Cohen, R. (Ed.) *Cambridge Survey of World Migration*. Cambridge, UK: Cambridge University Press, pp. 126–30.

Mohapatra, S., Ratha D. and Silwal, A. (2011) *Outlook for Remittance Flows 2011–13*. Migration and Development Brief 16. Migration and Remittance Unit. Washington, DC: The World Bank.

Moses, J.M. (2006) *International Migration: Globalization's Last Frontier*. London and New York: Zed Books.

Moya, J.C. (2007) Domestic Service in a Global Perspective: Gender, Migration, and Ethnic Niches. *Journal of Ethnic and Migration Studies* 33(4): 559–79.

Myers, N. (2002) Environmental Refugees: A Growing Phenomenon of the 21st Century. *Philosophical Transactions of the Royal Society of London. Series B: Biological Sciences* 357(1420): 609–13.

Navarro, M. (2013) Relying on Hotel Rooms for Thousands Uprooted by Hurricane Sandy. *New York Times*, March 29.

Ness, I. and Bellwood, P. (Eds) (2013) *The Encyclopedia of Global Human Migration*. Hoboken, NJ: Wiley.

Noiriel, G (1995) Italians and Poles in France. In Cohen, R. (Ed.) *Cambridge Survey of World Migration*. Cambridge, UK: Cambridge University Press, pp. 142–44.

Northrop, D. (1995) *Indentured Labor in the Age of Imperialism, 1834–1922*. Cambridge, UK: Cambridge University Press.

Okamura, J.Y. (1981) Situational Ethnicity. *Ethnic and Racial Studies* 4(4): 452–65.

O'Neil, K. (2003) *Brain Drain and Gain: The Case of Taiwan*. Available online at http://www.migrationinformation.org/feature/display.cfm?ID=155 (accessed April 12, 2013).

Park, R.E. (1930) Assimilation, Social. In Edwin, R., Seligman, A. and Johnson, A. (Eds) *Encyclopedia of the Social Sciences, Volume 2*. New York: Macmillan, p. 281.

Park, R.E. and Burgess, E.W. (1967) *The City*. Chicago, IL: University of Chicago Press.

Park, R.E., Burgess, E.W. and McKenzie, R. (1925) *The City*. Chicago: University of Chicago Press.

Parlow, M. (2007) A Localist's Case for Decentralizing Immigration Policy. *Denver University Law Review* 84: 1061.

Parreñas, R. (2001) *Servants of Globalization: Women, Migration, and Domestic Work*. Stanford, CA: Stanford University Press.

Passel, J.S. and Cohn, D. (2010) *U.S. Unauthorized Immigration Flows Are Down Sharply Since Mid-Decade*. Washington, DC: Pew Hispanic Center.

Peck, J. and Tickell, A. (2002) Neoliberalizing Space. *Antipode* 34(3): 380–404.

Pellegrino, A. (2001) Trends in Latin American Skilled Migration: "Brain Drain" or "Brain Exchange"? *International Migration* 39(5): 111–32.

Petras, J. and Veltmeyer, H. (2001) *Globalization Unmasked*. London: Zed Books.

Petrzela, N.M. (2013) Multiculturalism. In Ness, I. and Bellwood, P. (Eds) *The Encyclopedia of Global Human Migration*. Hoboken, NJ: Wiley. Available online at http://onlinelibrary.wiley.com/doi/10.1002/9781444351071.wbeghm377/abstract (accessed June 20, 2013).

Piore, M.J. (1979) *Birds of Passage: Migrant Labor and Industrial Societies*. Cambridge, UK: Cambridge University Press.

Plummer, K. (2010) *Sociology: The Basics*. Oxford, UK: Routledge.

Portes, A. (1997) Immigration Theory for a New Century: Some Problems and Opportunities. *International Migration Review* 31(4): 799–825.

Portes, A. and Bach, R. (1985) *Latin Journey*. Berkeley, CA: University of California Press.

Portes, A. and Jensen, L. (1987) What's an Ethnic Enclave. The Case for Conceptual Clarity. *American Sociological Review* 52(6): 768–70.

Portes, A. and Manning, R. (1986) The Immigrant Enclave: Theory and Empirical Examples. In Nagel, J. and Olzak, S. (Eds) *Competitive Ethnic Relations*. Orlando, FL: Academic Press.

Portes, A. and Rumbaut, R. (1996) *Immigrant America: A Portrait, Second Edition*. Berkeley, CA: University of California Press.

Portes, A. and Zhou, M. (1993) The New Second Generation: Segmented Assimilation and its Variants. *The Annals of the American Academy of Political and Social Science* 530(1): 74–96.

Powers, M. (2013) Assimilation, Integration and Incorporation. In Ness, I. and Bellwood, P. (Eds) *The Encyclopedia of Global Human Migration*. Hoboken, NJ: Wiley-Blackwell. Available online at http://onlinelibrary.wiley.com/doi/10.1002/9781444351071.wbeghm047/abstract (accessed May 10, 2013).

Price, M. and Benton-Short, L. (Eds) (2008) *Migrants to the Metropolis: The Rise of Immigrant Gateway Cities*. Syracuse, NY: Syracuse University Press.

Ratha, D. (2013) *The Impact of Remittances on Economic Growth and Poverty Reduction*. Washington, DC: Migration Policy Institute.

Ratha, D. and Shaw, W. (2007) South–South Migration and Remittances. Available online at http://www.migrationinformation.org/Feature/display.cfm?ID=641 (accessed January 3, 2013).

Ravenstein, E. (1885) The Laws of Migration. *Journal of the Statistical Society of London* 48(2): 167–235.

——(1889) The Laws of Migration. *Journal of the Royal Statistical Society of London* 52(2): 241–305.

Rediker, M. (2008) *The Slave Ship: A Human History*. New York: Penguin Books.

Rex, J. (1994) The Second Project of Ethnicity: Transnational Migrant Communities and Ethnic Minorities in Modern Multicultural Societies. *Innovation: The European Journal of Social Science Research* 7(3): 207–17.

Ridgley, J. (2008) Cities of Refuge: Immigration Enforcement, Police, and the Insurgent Genealogies of Citizenship in US Sanctuary Cities. *Urban Geography* 29(1): 53–77.

Ritzer, G. (2011) *Globalization the Essentials*. Malden, MA: Wiley-Blackwell.

Roberts, S. (2007) In shift: 40% of immigrants move directly to suburbs. *New York Times*, October 17, p. A22.

Rodriguez, C. (2007) *The Significance of the Local in Immigration Regulation*. New York Public Law and Legal Theory Working Papers, Paper 75. New York, NY: New York University Law School.

Rubinkam, M. (2011) Court Orders New Look at Pa. City Immigration Law. *The Seattle Times*, June 6.

Sanders, J. and Nee, V. (1987) Limits of Ethnic Solidarity in the Enclave Economy. *American Sociological Review* 52(6): 745–73.

Sassen, S. (1988) *The Mobility of Labor and Capital. A Study in International Investment and Labor Flow*. Cambridge, UK: Cambridge University Press.

——(1991) *The Global City: New York, London, Tokyo*. Princeton, NJ: Princeton University Press.

Saunders, D. (2010) *Arrival Cities: How the Largest Migration in History is Reshaping Our World*. New York: Pantheon.

Scally, R. (1995) The Irish and the "Famine Exodus" of 1847. In Cohen, R. (Ed.) *Cambridge Survey of World Migration*. Cambridge, UK: Cambridge University Press, pp. 80–84.

Schiff, M. and Özden C. (Eds) (2006) *International Migration, Remittances and Brain Drain*. New York: Palgrave Macmillan.

Scott, A.J. and Soja, E.W. (1996) *The City: Los Angeles and Urban Theory at the End of the Twentieth Century*. Berkeley, CA: University of California Press.

Sezneva, O. (2013) Ethnicity Theory for Migration Research. In Ness, I. and Bellwood, P. (Eds) *The Encyclopedia of Global Human Migration*. Hoboken, NJ: Wiley-Blackwell. Available online at http://onlinelibrary.wiley.com/doi/10.1002/9781444351071.wbeghm203/abstract (accessed June 5, 2013).

Sheffer, G. (1986) *Modern Diasporas in International Politics*. New York: Palgrave Macmillan.

Shenon, P. (2008) *The Commission: The Uncensored History of the 9/11 Investigation*. New York, NY: Hachette Book Group.

Shimpo, M. (1995) Indentured Servants from Japan. In Cohen, R. (Ed.) *Cambridge Survey of World Migration*. Cambridge, UK: Cambridge University Press, pp. 48–50.

Short, J.R. and Kim, Y.H. (1999) *Globalization and the City*. Upper Saddle River, NJ: Prentice Hall.

Singer, A., Hardwick, S.W. and Brettell, C.B. (2008) *Twenty-First Century Gateways: Immigrant Incorporation in Suburban America*. Washington, DC: Brookings Institution Press.

Singer, B. (2013) *The Americanization of France: Searching for Happiness after the Algerian War*. Lanhan, MD: Rowman & Littlefield Publishers.

Smith, J.P. and Edmonston, B. (1997) *The New Americans: Economic, Demographic and Fiscal Effects of Immigration*. Washington, DC: National Academies Press.

Solimano, A. (2010) *International Migration in the Age of Crisis and Globalization*. Cambridge, UK: Cambridge University Press.

Solomon, S., Qin, D., Manning, M., Chen, Z., Marquis, M., Averyt, K.B., Tignor, M. and Miller, H.L. (2007) *Contribution of Working Group I to the Fourth Assessment Report of the Intergovernmental Panel on Climate Change*. Cambridge, UK: Cambridge University Press.

Spencer, R.G. (1997) *British Immigration Policy Since 1939: The Making of Multi-Racial Britain*. New York: Routledge.

Stalker, P. (2000) *Workers Without Frontiers: The Impact of Globalization on International Migration*. Boulder, CO: Lynne Rienner Publishers.

——(2001) *The No-Nonsense Guide to International Migration*. New York: Verso, 2001.

Stern, N. (2007) *The Economics of Climate Change: The Stern Review*. Cambridge, UK: Cambridge University Press.

Tan, C. (2007) Liberalization. In Scholte, J.A. and Robertson, R. (Eds) *Encyclopedia of Globalization*. New York: MTM Publishing, pp. 735–9.

Taylor, C. (1994) The Politics of Recognition. In Taylor, C. and Gutmann, A. (Eds) *Multiculturalism*. Princeton, NJ: Princeton University Press, pp. 25–74.

Taylor, P.J. (2004) *World City Network: A Global Urban Analysis*. New York: Routledge.

Terry, D.F. and Wilson, S.R. (Eds) (2005) *Beyond Small Change: Making Migrant Remittances Count*. Washington, DC: Inter-American Development Bank.

Thiara, R.K. (1995) Indian Indentured Workers in Mauritius, Natal and Fiji. In Cohen, R. (Ed.) *Cambridge Survey of World Migration*. Cambridge, UK: Cambridge University Press, pp. 63–8.

Tribalat, M. (1995) *Faire France: une enquête sur les immigrés et leurs enfants*. Paris: La Découverte.

United Nations (1997) International Migration and Development: The Concise Report. Population Division. *Population Newsletter* 63, New York: United Nations.

——(2012) *World Urbanization Prospects: The 2011 Revision*. New York: Population Division of the Department of Economic and Social Affairs, United Nations.

United Nations Development Programme (2009) *Human Development Report Overcoming Barriers: Human Mobility and Development*. New York, NY: United Nations Development Programme.

U.S. Census Bureau (2011) Estimate of Place of Birth by Year of Entry by Citizenship Status for the Foreign-born Population. In *American Community Survey*. Washington, DC.

U.S. Department of State (2007) Trafficking in Persons Report 2007. Washington, DC: US Government Printing Office. Available online at http://www.state.gov/documents/organization/82902.pdf (accessed June 19, 2013).

U.S. Environmental Protection Agency (2010) *Climate Change Indicators in the United States*. Washington, DC: United States Environmental Protection Agency Climate Change Division.

Varsanyi, M.W. (2008) Rescaling the "Alien", Rescaling Personhood: Neoliberalism, Immigration and the State. *Annals of the Association of American Geographers* 98(4): 877–96.

Vicino, T.J. (2013) *Suburban Crossroads: The Fight for Local Control of Immigration Policy*. Lanham, MD: Lexington Books.

Vicino, T.J., Hanlon, B. and Short, J.R. (2007) Megalopolis 50 Years On: The Transformation of a City Region. *International Journal of Urban and Regional Research* 31(2): 344–67.

Vogel, D. (2009) Comparative Policy Brief: The Size of Irregular Migration. Size and Development of Irregular Migration to the EU Clandestino Research Project, The European Commission. October 2009. Available online at http://clandestino.eliamep.gr/wpcontent/uploads/2009/12/clandestino_poli cy_brief_comparative_size-of-irregular-migration.pdf (accessed July 14, 2013).

Vogel, R. (2013) Marxist Theories of Migration. In Ness, I. and Bellwood, P. (Eds) *Encyclopedia of Global Human Migration*. Hoboken, NJ: Wiley-Blackwell Publishers. Available online at http://onlinelibrary.wiley.com/doi/10.1002/ 9781444351071.wbeghm354/abstract (accessed June 5, 2013).

Waldinger, R. (1993) The Ethnic Enclave Debate Revisited. *International Journal of Urban and Regional Research* 17(3): 444–52.

——(1996) *Still the Promised City? African-Americans and New Immigrants in Postindustrial New York*. Cambridge, MA: Harvard University Press.

Waldinger, R. and Bozorgmehr, M. (Eds) (1996) *Ethnic Los Angeles*. New York: Russell Sage.

Wallerstein, I. (1974) *The Modern World-System, Vol. I: Capitalist Agriculture and the Origins of the European World-Economy in the Sixteenth Century*. New York: Academic Press.

Warren, R. and Warren, J.R. (2013) Unauthorized Immigration to the United States: Annual Estimates and Components of Change, by State, 1990 to 2010. *International Migration Review* 47(2): 296–329.

Waters, M.C. (1999) *Black Identities: West Indian Immigrant Dreams and American Realities*. New York: Russell Sage Foundation.

Weil, P. (1991) Immigration and the Rise of Racism in France: The Contradictions of Mitterrand's policies. *French Society and Politics* 9: 3–4.

Weiner, M. (1995) *The Global Migration. Crisis: Challenge to States and to Human Rights*. New York: Harper Collins College Publishers.

Wilson, K. and Portes, A. (1980) Immigrant Enclaves: An Analysis of Labor Market Experiences of Cubans in Miami. *American Journal of Sociology* 86(2): 295–319.

World Bank (2013) *Migration and Development Brief*. Washington, DC: Migration and Remittances Unit, Development Prospects Group, April 13.

Yue, A. (2008) Same-Sex Migration in Australia: From Interdependency to Intimacy. *GLQ: A Journal of Lesbian and Gay Studies* 14(2–3): 239–62.

——(2013) Lesbian, Gay, Bisexual, Transgender (LGBT) Migration. In Ness, I. and Bellwood, P. (Eds) *The Encyclopedia of Global Human Migration*. Hoboken, NJ: Wiley-Blackwell. Accessed online July 16, 2013.

Zheng, N. (2007) *Chinese Green Card System Continues to Evolve*. Available online at http://www.china.org.cn/english/LivinginChina/221738.htm (accessed April 20, 2011).

Zhou, M. (1992) *Chinatown: The Socioeconomic Potential of an Ethnic Enclave*. Philadelphia, PA: Temple University Press.

——(2013) Ethnic Enclaves and Niches. In Ness, I. and Bellwood, P. (Eds) *The Encyclopedia of Global Human Migration*. Hoboken, NJ: John Wiley & Sons.

Zhou, M. and Cai, G. (2007) Chinese Language Media and the Ethnic Enclave Economy in the United States. In Fong, E. and Luk, C. (Eds) *Chinese Ethnic Business: Global and Local Perspectives*. New York: Routledge, pp. 21–34.

Zolberg, A.R. (2001) Introduction: Beyond the Crisis. In Zolberg, A.R. and Benda, P.M. (Eds) *Global Migrants, Global Refugees: Problems and Solutions*. New York/Oxford: Berghahn, pp. 1–16.

INDEX

Taylor & Francis
eBooks
FOR LIBRARIES
ORDER YOUR FREE 30 DAY INSTITUTIONAL TRIAL TODAY!
Over 22,000 eBook titles in the Humanities, Social Sciences, STM and Law from some of the world's leading imprints.
Choose from a range of subject packages or create your own!
Benefits for you
Free MARC records
COUNTER-compliant usage statistics
Flexible purchase and pricing options
Benefits for your user
Off-site, anytime access via Athens or referring URL
Print or copy pages or chapters
Full content search
Bookmark, highlight and annotate text
Access to thousands of pages of quality research at the click of a button
For more information, pricing enquiries or to order a free trial, contact your local online sales team.
UK and Rest of World: online.sales@tandf.co.uk
US, Canada and Latin America: e-reference@taylorandfrancis.com
www.ebooksubscriptions.com
ALPSP Award for BEST eBOOK PUBLISHER 2009 Finalist
Taylor & Francis eBooks
Taylor & Francis Group
A flexible and dynamic resource for teaching, learning and research.